Endorsed by
University of Cambridge International Examinations

Biology
IGCSE

Revision
Guide

Ron Pickering

OXFORD

OXFORD

UNIVERSITY PRESS

Great Clarendon Street, Oxford OX2 6DP

Oxford University Press is a department of the University of Oxford.

It furthers the University's objective of excellence in research, scholarship, and education by publishing worldwide in

Oxford   New York

Auckland   Cape Town   Dar es Salaam   Hong Kong   Karachi
Kuala Lumpur   Madrid   Melbourne   Mexico City   Nairobi
New Delhi   Shanghai   Taipei   Toronto

With offices in

Argentina   Austria   Brazil   Chile   Czech Republic   France   Greece
Guatemala   Hungary   Italy   Japan   Poland   Portugal   Singapore
South Korea   Switzerland   Thailand   Turkey   Ukraine   Vietnam

British Library Cataloguing in Publication Data

Data available

ISBN: 978-0-19-915265-0
10 9 8 7 6 5

Printed in United Kingdom by Bell and Bain Ltd., Glasgow

Paper used in the production of this book is a natural, recyclable product made from wood grown in sustainable forests. The manufacturing process conforms to the environmental regulations to the country of origin.

CIE past question paper examination material reproduced by permission of the University of Cambridge Local Examinations Syndicate.

The University of Cambridge Local Examinations Syndicate bears no responsibility for the example answers to questions taken from its past question papers which are contained in this publication.

Cover photo: IGCSE BIOLOGY © blickwinkel/Alamy

Contents

Answers provided on our website: www.oxfordsecondary.co.uk/biologyrg

Introduction

This book has been written to help the reader manage both the demands of content and skills required by the Cambridge International Examinations IGCSE Specification. The content is presented as a series of annotated diagrams, each carefully matched to the CIE specification and offering factual material in easily-manageable "chunks". The skills are developed in two ways – as a set of guidelines for revision and for practical work, and as a series of questions of exactly the type used for the CIE examinations. In addition, each section contains a vocabulary exercise – either in crossword or "fill the gap" format. These exercises will prove invaluable for those students for whom English is not the first language.

The material in the Study Guide is arranged to closely follow the sequence of topics in the Complete Biology for IGCSE textbook: students will be able to use the Study Guide to support their learning from the Complete Biology textbook, or as an independent revision guide.

Chapter 1:
Four steps to success in examinations

If you have already read about the brain you might remember that an important function of the cerebral hemispheres is to act as an **integration centre** – input of information is compared with previous experience and an appropriate action is taken. As you prepare for an examination you will be inputting factual material and skills and you will be hoping that, when you're faced with examination papers, you will be able to make the correct responses! The effort you make in **revision** and your willingness to **listen to advice on techniques** will greatly affect your likelihood of success, as outlined here.

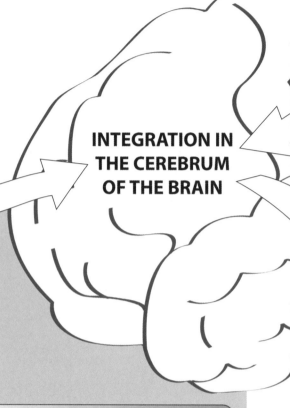

INTEGRATION IN THE CEREBRUM OF THE BRAIN

1 | REVISION

There is no one method of revising that works for everyone. It is therefore important to discover the approach that suits you best. The following rules may serve as general guidelines.

Leaving everything until the last minute reduces your chances of success. Work will become more stressful, which will reduce your concentration. There are very few people who can revise everything 'the night before' and still do well in an examination the next day.

PLAN YOUR REVISION TIMETABLE

You need to plan your revision timetable some weeks before the examination and make sure that your time is shared suitably between all your subjects.
Once you have done this, follow it – don't be side-tracked. Stick your timetable somewhere prominent where you will keep seeing it – or better still put several around your home!

RELAX

Concentrated revision is very hard work. It is as important to give yourself time to relax as it is to work. Build some leisure time into your revision timetable.

GIVE YOURSELF A BREAK

When you are working, work for about an hour and then take a short tea or coffee break for 15 to 20 minutes. Then go back to another productive revision period.

KEEP TRACK

Use checklists and the relevant examination board specification to keep track of your progress. Mark off topics you have revised and feel confident with. Concentrate your revision on things you are less happy with.

MAKE SHORT NOTES, USE COLOURS

Revision is often more effective when you do something active rather than simply reading material. As you read through your notes and textbooks make brief notes on key ideas. If this book is your own property you could highlight the parts of pages that are relevant to the specification you are following. Concentrate on understanding the ideas rather than just memorizing the facts.

PRACTISE ANSWERING QUESTIONS

As you finish each topic, try answering some questions. There are some in this book to help you (see Chapter 2). You should also use questions from past papers. At first you may need to refer to notes or textbooks. As you gain confidence you will be able to attempt questions unaided, just as you will in the exam.

2 KNOW WHAT TO DO

LEARN THE KEY WORDS

Name: the answer is usually a technical term (mitochondrion, for example) consisting of no more than a few words. **State** is very similar, although the answer may be a phrase or sentence. **Name** and **state** don't need anything added, i.e. there's no need for explanation.

Define: the answer is a formal meaning of a particular term, i.e. 'what is it?'
What is meant by...? is often used instead of **define**.

List: you need to write down a number of points (each may only be a single word) with no need for explanation.

Describe: your answer will simply say what is happening in a situation shown in the question, e.g. 'the temperature increased by 25°C' – There is no need for explanation.

Suggest: you will need to use your knowledge and understanding of biological topics to explain an effect that may be new to you. You might use a principle of enzyme action to suggest what's happening in an industrial process, for example. There may be more than one acceptable answer to this type of question.

Explain: the answer will be in extended prose, i.e. in the form of complete sentences. You will need to use your knowledge and understanding of biological topics to write more about a statement that has been made in the question or earlier in your answer. Questions like this often ask you to **state and explain.**

Calculate: a numerical answer is to be obtained, usually from data given in the question. Remember to:
- give your answer to the correct number of significant figures, usually two or three
- give the correct unit
- show your working.

3

IN THE EXAMINATION

- Check that you have the correct question paper! There are many options in some specifications, so make sure that you have the paper you were expecting.
- Read through the whole paper before beginning. Select the questions you are most comfortable with – there is no rule that says you must answer the questions in the order they are printed!
- Read the question carefully – identify the key word (why not underline it?).
- Don't give up if you can't answer part of a question. The next part may be easier and may provide a clue to what you should do in the part you find difficult.
- Check the number of marks allocated to each section of the question.
- Read data from tables and graphs carefully. Take note of column headings, labels on axes, scales, and units used.
- Keep an eye on the clock – perhaps check your timing after you've finished 50 per cent of the paper.
- Use any 'left over' time wisely. Don't just sit there and gaze around the room. Check that you haven't missed out any sections (or whole questions! Many students forget to look at the back page of an exam paper!). Repeat calculations to make sure that you haven't made an arithmetical error.

4 *Success!*

On the following pages some sample examination questions are answered. As you study them try to notice the following key points:

- what are you told to do and has the answer done it? Key words to look for are:
  state; describe; explain; suggest; predict; calculate
- the number of marks offered must be matched by the number of points made
- can you recognize what part of the syllabus is covered by the question because this will guide you to the likely answer
- some questions cover complicated material but actually contain most of the information which you need to gain full marks!

1. The graph below shows the volume of air in the lungs of a person measured over a period of time.

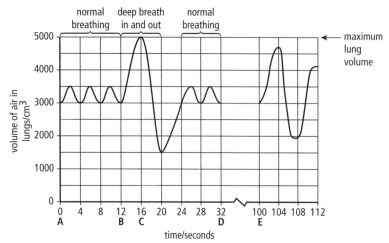

a. i. With reference to the graph calculate, in breaths per minute, the rate of normal breathing between **A** and **B**. Show your working.

> Don't forget! You will get a mark for correct working even if you make a mistake with the answer.

3 complete breaths in 12 seconds

So   3 x $\frac{60}{12}$ breaths in 60 seconds (ie 1 minute)

= 3 x 5 = 15   ....15.... breaths per minute (2)

> Don't forget the units!

ii. State the volume of air remaining in the lungs after the deep breath out.

1500 cm³

(1)

iii. Explain how the intercostal muscles are involved in breathing from time **B** to time **C**.

> **Explain** is more than **describe**. Can your answer begin with the word 'because'?

Volume increases because external intercostal muscles

contract (1 mark) and lift the rib edge upwards

and outwards (1 mark)

(2)

At time D, the person performed one minute of vigorous exercise.

b. i. On the graph starting at time **E**, continue the graph to show the person's brething pattern after this exercise.

(2)

> Notice this – many students don't!

> Person should be recovering from deeper, faster breathing.

ii. Explain why the breathing pattern changes after a period of exercise.

Because

- Less oxygen required/less CO_2 to be excreted
- Fewer, shallower breaths will be enough

(3)

(Total 10 marks)

CIE 0610 June '98 Paper 3 Q1

2. a.   State **three** normal functions of a root.

1. To anchor the plant in the soil

2. To absorb water from the soil

3. To absorb mineral ions from the soil

(3)

The diagram below shows a mangrove tree growing in a swamp.

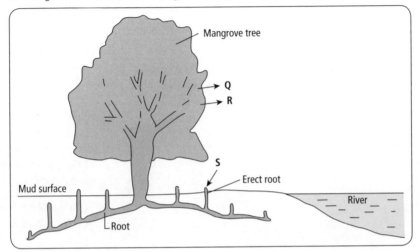

The roots of the mangrove are specially modified to overcome the fact that air spaces in the soil are always filled with water. In other respects, the roots are normal in structure and function.

Arrows **Q**, **R** and **S** represent the movement of gases into and out of the tree during the day.

So light is available for photosynthesis

b. i.   Name gases **Q** and **R** and, for each gas, state the process in the tree which produces it.

Gas **Q** Oxygen   process photosynthesis

Gas **R** Carbon dioxide   process respiration

(2)

You need **both** answers to gain the mark–the gas must match the process

ii.   Name gas **S** and state in which process it is used in the tree.

Gas **S** Oxygen   process respiration

(1)

Roots don't photosynthesise

c.   Name the unusual response being shown by the erect roots of the tree.

Positive phototropism (or negative geotropism)

(1)

Active transport is a process which occurs in most plant roots.

d.   Suggest why mangrove roots may have difficulty in carrying out this process.

• Active transport requires energy from respiration

• Roots can't obtain much oxygen for aerobic respiration

(2)

Your answer should refer to some biological knowledge

(Total 9 marks)

CIE 0610 June '98 Paper 3 Q2

3. The table below shows how three alleles control the ABO human blood group system.

ALLELE	EFFECT
I^A	Causes the production of antigen A on red blood cells
I^B	Causes the production of antigen B on red blood cells
I^O	Does not cause the production of antigens on red blood cells

When I^A and I^B are inherited together they are equal in their effect. When I^O is inherited ⟶ with either I^A or I^B it is recessive.

Important information which will be used later in the question

a. Complete the diagram below by filling in the blank circles to show the possible inheritance of the ABO blood group system in Mr and Mrs Shah's children.

*A clear **instruction** – no need for any **explanation** here*

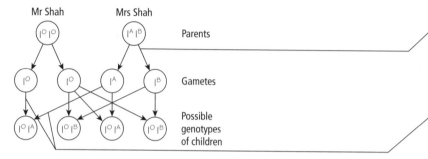

At meiosis (reduction division) only one of each pair of alleles from the diploid parent can be passed to any single haploid gamete

At fertilisation two haploid gametes fuse to form a diploid zygote

b. When I^A and I^B are inherited together they both show their effects in the phenotype. Name the term used to describe this.

Co-dominance .... (1)

Important! No call for any explanation

Define the term phenotype

The total of the features of characteristics of an organism .... (1)

c. i. What is the probability of the eldest child being female?

50% or ½ or 1 in 2 .... (1)

Sex is determined by the X or Y chromosome carried in the male gamete. There are equal numbers of 'X' and 'Y' gametes.

ii. What is the probability of the eldest child being Female with blood group B?

½ (being female) x ½ (being B) = ¼ or 25% or 1 in 4 .... (1)

*N.B. **not** 1:4, as this means 1 in 5 or 20%*

d. Sometimes a patient needs a blood transfusion. It is very important that the blood given to the patient does not contain any antigen which is not in the patient's own red blood cells. Which parent can donate blood safely to any of the possible children?

Mr Shah (Since I^O blood has no damaging antigens) .... (1)

(Total 8 marks)

4. a. Select the correct term from the list below and write it in the box next to its description.

— a very clear instruction.

allele   dominant   gene   genotype   heterozygous

homezygous   phenotype   recessive

DESCRIPTION	TERM
a form of a gene that always has its effect when it is present	dominant
a form of a gene that codes for one of a pair of contrasting features	allele
an organism having two different forms of a gene for a particular feature	heterozygous
that alleles that an organism has in its chromosome	genotype

(4)

b. Two red flowered plants were crossed. The seeds produced were germinated and grew into 62 white flowered plants and 188 red flowered plants.

   i. Which flower colour is controlled by the recessive form of the gene?
   white (the 'l' in the 3:l)

   (1)

   ii. Using the symbols **R** and **r**, construct a genetic diagram to explain the results of this cross.

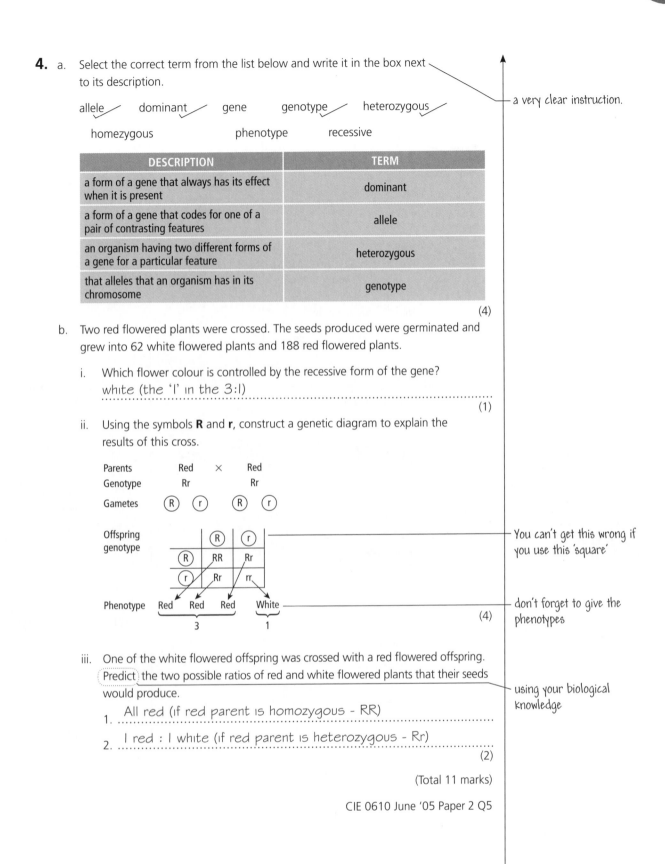

Parents   Red   ×   Red
Genotype   Rr   Rr
Gametes   (R) (r)   (R) (r)

Offspring genotype
	(R)	(r)
(R)	RR	Rr
(r)	Rr	rr

— You can't get this wrong if you use this 'square'

Phenotype   Red Red Red   White
              └── 3 ──┘   └ 1 ┘

— don't forget to give the phenotypes

(4)

   iii. One of the white flowered offspring was crossed with a red flowered offspring. Predict the two possible ratios of red and white flowered plants that their seeds would produce.

   1. All red (if red parent is homozygous - RR)

   2. l red : l white (if red parent is heterozygous - Rr)

— using your biological knowledge

(2)

(Total 11 marks)

CIE 0610 June '05 Paper 2 Q5

5. The diagram below shows a food web from the Antarctic.

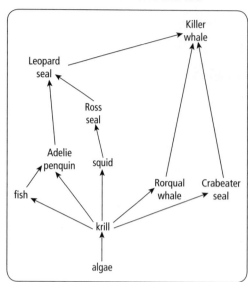

a.   i.   ⟨State⟩ the original source of energy for this food web.   no need for explanation
         Sunlight
         ....
                                                                                                    (1)
     ii. ⟨Name⟩ an organism in this food web that is both a secondary and a tertiary
         consumer
         Adelie penguin ——————————————————————————   eats krill (secondary)
         ....   and fish (tertiary)
                                                                                                    (1)
b.   Write in the names of organisms to form a complete food chain.

Killer whale

↑

Leopard seal

↑   count the number of steps

Adelie penguin

↑

Krill

↑

Algae

                                                                                          must begin with the
                                                                        (1)   producer

c.   There is concern that pollution of the environment may change the breeding
     grounds of the Adelie penguin.

                                                                                          two pieces of information
     ⟨State and explain⟩ the effect his might have on the populations of the Leopard seal   needed – can you link them
     and the Ross seal.   with the word 'because'?

     Leopard seal ...Population could fall because there would.....
                     be less food for them (so they would
                     ....
                     breed less successfully)
                     ....

     Ross seal ...Population could rise because there would....
                   be fewer leopard seals to act as predators
                   ....
                                                                                    (4)
                                                                        (Total 7 marks)

CIE 0610 June '05 Paper 2 Q6 ▼

6. The graph below shows the heart rate and the cardiac output. The cardiac output is the volume of blood pumped out of the heart each minute.

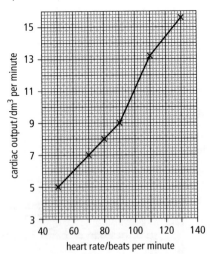

a. i.   What is the cardiac output at a heart rate of 100 beats per minute?_____

11 dm³ per minute ....

read directly (and accurately!) from the graph

don't forget the units!

(1)

ii. Determine the increase in cardiac ouput when the heart rate increases from 70 to 90 beats per minute

7⌐   ⌐9

(9 – 7) = 2 .... dm³ per minute   (1)

means the same as 'calculate' (or 'work out')

iii. Determine the increase in cardiac output when the heart rate increases from 100 to 120 beats per minute.

11⌐   ⌐14.4

(14.4 – 11) = 3.4 .... dm³ per minute   (1)

b. i.   Which chamber of the heart pumps blood into the aorta?_____

left ventricle ....

(1)

ii. The upper and lower chambers on each side of the heart are separated by valves.

State the function of these values.

to prevent the back flow of blood (from ventricle to atrium) ....

....

(1)

Simple recall of facts, you just have to know when to present them

CIE 0610 November '05 Paper 2 Q7

Chapter 3:
Design an experiment

An experiment is designed to **test the validity of a hypothesis** and involves the **collection of data**.

e.g. light intensity affects the rate of photosynthesis

using appropriate apparatus and instruments

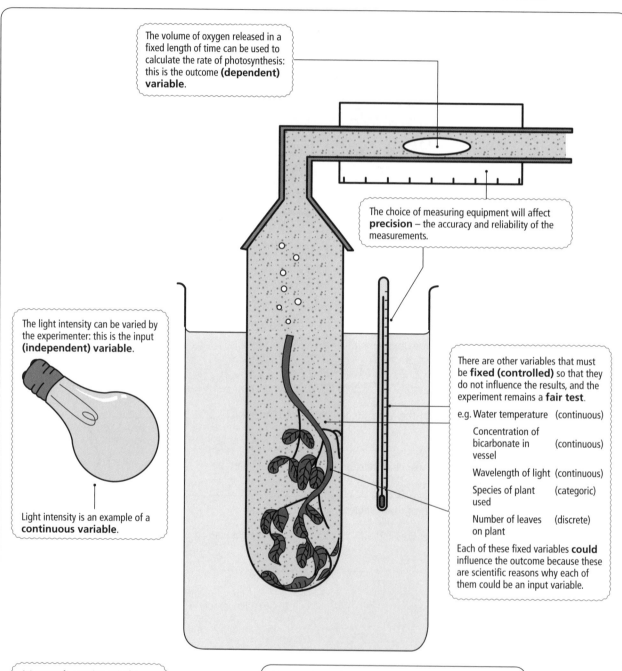

The volume of oxygen released in a fixed length of time can be used to calculate the rate of photosynthesis: this is the outcome **(dependent) variable**.

The choice of measuring equipment will affect **precision** – the accuracy and reliability of the measurements.

The light intensity can be varied by the experimenter: this is the input **(independent) variable**.

Light intensity is an example of a **continuous variable**.

There are other variables that must be **fixed (controlled)** so that they do not influence the results, and the experiment remains a **fair test**.

e.g. Water temperature (continuous)

Concentration of bicarbonate in vessel (continuous)

Wavelength of light (continuous)

Species of plant used (categoric)

Number of leaves on plant (discrete)

Each of these fixed variables **could** influence the outcome because these are scientific reasons why each of them could be an input variable.

A 'control' experiment is the same in every respect, **except** the input variable is not changed but is kept constant.

A control allows confirmation that **no unknown variable** is responsible for any observed changes in the responding variable; it helps to make the experiment a **fair test**.

A '**repeat**' is performed when the experimenter suspects that misleading data has been obtained through 'operator error'.

'**Means**' of a series of results minimize the influence of any single result, and therefore reduce the effect of any 'rogue' or anomalous data. The use of means improves the RELIABILITY of data – how confident you are about the observations or measurments.

Chapter 4:
Dealing with data

May involve a number of steps

1. Organization of the **raw data** (the information which you actually collect during your investigation).

2. Manipulation of the data (converting your measurements into another form).

and

3. Representation of the data in **graphical** or other form.

Step 1 and 2 usually involve **preparation of a table of results**

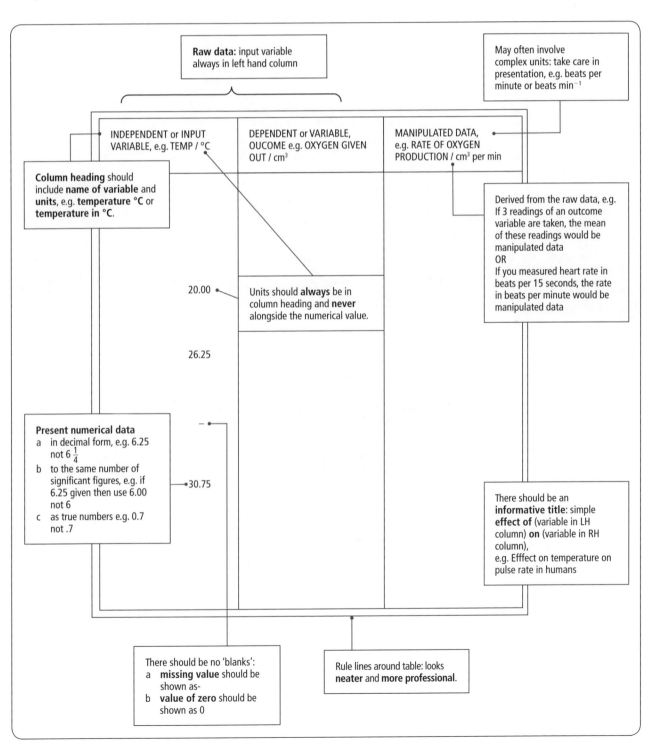

Chapter 5:
Graphical representation

A graph is a visual presentation of data and may help to make the relationship between variables more obvious, i.e. allow the experimenter to **draw a conclusion**.

For example

AIR TEMPERATURE / °C	BODY TEMPERATURE OF REPTILE / °C
20	19.4
25	25.9
30	30.4
35	35.1
40	40.1
45	44.8

isn't as helpful as

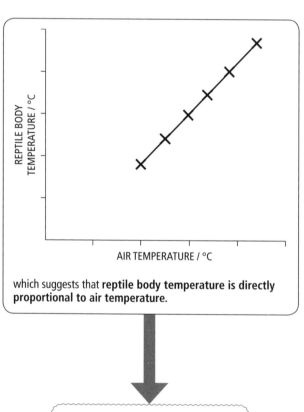

which suggests that **reptile body temperature is directly proportional to air temperature.**

Validity – how much confidence can be placed in the conclusion. Data is only valid if there is **only one input variable**. Can be affected by the **range** and **reliability** of measurements.

There should be an **informative title:** for example **Effect of** (variable on *x* axis) **on** (variable on *y* axis).

A graph may be produced from a table of data **following certain rules**

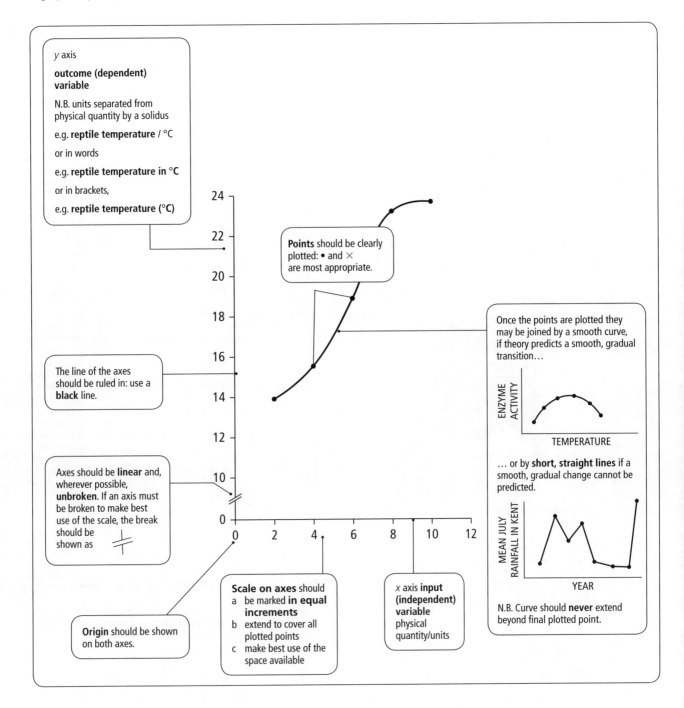

y axis

outcome (dependent) variable

N.B. units separated from physical quantity by a solidus

e.g. **reptile temperature / °C**

or in words

e.g. **reptile temperature in °C**

or in brackets,

e.g. **reptile temperature (°C)**

Points should be clearly plotted: • and ×
are most appropriate.

The line of the axes should be ruled in: use a **black** line.

Once the points are plotted they may be joined by a smooth curve, if theory predicts a smooth, gradual transition…

ENZYME ACTIVITY

TEMPERATURE

Axes should be **linear** and, wherever possible, **unbroken**. If an axis must be broken to make best use of the scale, the break should be shown as

… or by **short, straight lines** if a smooth, gradual change cannot be predicted.

MEAN JULY RAINFALL IN KENT

YEAR

Scale on axes should
a be marked **in equal increments**
b extend to cover all plotted points
c make best use of the space available

x axis **input (independent) variable** physical quantity/units

N.B. Curve should **never** extend beyond final plotted point.

Origin should be shown on both axes.

Chapter 6:
The variety of life

LIVING THINGS: THEIR CHARACTERISTICS

The biosphere is made up of living and non-living things. Plants and animals are living things because they have these characteristics:

RESPIRATION: the chemical reactions that break down nutrient molecules in living cells to release energy.
Food + oxygen ⟶
    energy + water + carbon dioxide

Respiration is the most important process. No organism is alive unless it can release energy by respiration.

GROWTH: a permanent increase in size and dry mass by an increase in cell number or cell size or both.

MOVEMENT: an action by an organism or part of an organism causing a change of position or place.

EXCRETION: removal from organisms of toxic materials, the waste products of metabolism (chemical reactions in cells including respiration) and substances in excess of requirements.

NUTRITION: the process of taking in of nutrients which are organic substances and mineral ions, containing raw materials or energy for growth and tissue repair, absorbing and assimilating them.

SENSITIVITY: the ability to detect or sense changes in the environment (stimuli) and to make responses.

REPRODUCTION: the processes that make more of the same kind of organism.

THE VARIETY OF LIVING ORGANISMS

A LIVING ORGANISM will show some or all of the characteristics of life – RESPIRATION is a particularly useful one to look for, as all living organisms need energy all of the time! You can find out which type of living organism it is by observing it carefully and then answering the following questions:

1	Is the organism made of a single cell?	YES	Go to question 2
		NO	Go to question 3
2	Do the cells have no clear nucleus?	YES	It's a **bacterium**
		NO	It's a **protoctistan**
3	Do the cells have no cell wall?	YES	It's an **animal** – go to question 5
4	Do the cells have an obvious cell wall?	YES	Go to question 14
5	Does the animal have a backbone and internal skeleton?	YES	Go to question 6
		NO	Go to question 10
6	Does the animal have a smooth, moist skin?	YES	It's an amphibian
		NO	Go to question 7
7	Are there scales on the skin?	YES	Go to question 8
		NO	Go to question 9
8	Is the skin dry?	YES	It's a reptile
		NO	It's a fish
9	Does it have feathers, wings and a beak?	YES	It's a bird
		NO	It's a mammal
10	Does it have a segmented body?	YES	Go to question 11
		NO	It's a mollusc
11	Does it have jointed limbs?	YES	Go to question 12
		NO	It's an annelid
12	Does it have three pairs of legs?	YES	It's an insect
		NO	Go to question 13
13	Does it have four pairs of legs?	YES	It's a spider
		NO	It's a crustacean
14	Do the cells contain chlorophyll in chloroplasts?	YES	It's a **plant** – go to question 15
		NO	It's a **fungus**
15	Does it have stem, leaves and roots?	YES	Go to question 17
		NO	Go to question 16
16	Does it have stems and leaves but no roots?	YES	It's a moss
		NO	It's an alga
17	Does it produce spores?	YES	It's a fern
	Does it produce seeds?	YES	It's a flowering plant (Angiosperm)

NAMING THE ORGANISMS

Linnaeus was a scientist who believed he could put every organism into a group (the science of TAXONOMY) and give every organism a name (the science of NOMENCLATURE).

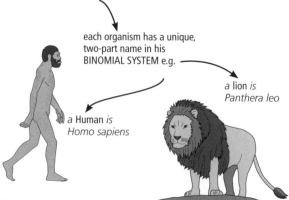

each organism has a unique, two-part name in his BINOMIAL SYSTEM e.g.

a lion is Panthera leo

a Human is Homo sapiens

THE DIVERSITY OF LIFE

KINGDOM	GROUP	
Plant	Monocotyledons e.g. grasses	• only one 'seed leaf' • long thin leaves, with parallel veins
	Dicotyledons e.g. trees	• two 'seed leaves' • broad leaves with a network of veins
Animal: Arthropods (have a hard exoskeleton and jointed limbs)	Insects	• three body parts – head, thorax and abdomen • three pairs of legs • usually 1 or 2 pairs of wings
	Crustaceans	• many segments, usually each with legs, claws or feelers • breathe via gills, as live in water • exoskeleton particularly hard, for protection
	Arachnids	• two body parts, eight legs and no wings • all have piercing jaws since all are predators
	Myriapods	• long, thin body with many segments for moving easily through soil and leaf litter • antennae as sense organs in dark habitats
	Annelids (true worms)	• many segments on long body • body covered with mucus to conserve water • many are hermaphrodite • usually have chetae (bristles) for movement
	Nematodes	• no segments - long cylindrical body
	Molluscs	• have a soft body, with a muscular foot used for movement • often have eyes on retractable tentacles • soft body is often enclosed in a shell made of calcium carbonate (protection from predators/drying out)

HINT! It is easier to remember these groups if you use examples from near where you live!

VERTEBRATES (ANIMALS WITH BACKBONES)

CLASS	EXTERNAL FEATURES	OTHER FEATURES
Fish (all aquatic)	• Scales • Fins • Eyes and lateral line	• Jelly-covered eggs; usually use external fertilisation • Ectothermic • Gills for gas exchange
Amphibians (always breed in water)	• Moist skin • Four limbs • Eyes and ears	• Jelly-covered eggs; external fertilisation • Ectothermic • Lungs/skin for gas exchange
Reptiles (lay eggs on land)	• Dry, scaly skin • Four limbs (not in snakes) • Eyes and ears	• Soft-shelled eggs; internal fertilisation • Ectothermic • Lungs for gas exchange
Birds (very few are aquatic)	• Feathers (scales on legs) • Two wings, two legs • Eyes and ears	• Hard-shelled eggs; internal fertilisation • Endothermic • Lungs for gas exchange
Mammals (very few are aquatic)	• Fur or hair • Four limbs • Eyes and ears • Nipples	• Live young (a few lay eggs) • Endothermic • Lungs for gas exchange • Feed young with milk from mammary glands

You could be asked to directly describe these in exam questions

You could use these features in questions on other topics

1. Select from the list the name of the group of animals that best fits each description.

Write your choice in the table

Arachnid   bird   crustacean   insect
        mammal   mollusc   nematode

DESCRIPTION OF ANIMAL	GROUP
a hard exoskeleton and more than 4 pairs of legs	arachnid
a hard shell and a slimy muscular foot	mollusc
one pair of wings and a beak	bird
one pair of wings and has skin covered with fur	mammal
two pairs of wings and one pair of antennae	insect

2. a. A biology student working in a museum mixed up a series of specimens. The specimens are named in this list:

AMPHIBIAN, EARTHWORM, BACTERIUM, BIRD, MAMMAL, FISH, REPTILE, FUNGUS, MOSS, FERN, PROTOCTISTAN, VIRUS, FLOWERING PLANT, INSECT, PARASITE

Choose organisms from this list to match the following descriptions:

Mammal i.   body covered with fur, has mammary glands
fungus ii.   cells with cell wall but no chlorophyll; reproduce by spores
reptile iii.   body covered with dry scales
insect iv.   body has three regions – head, thorax and abdomen
bacterium v.   single cell, without definite nucleus
virus vi.   can only reproduce within another living cell; few genes inside protein coat
bird vii.   feathers, beak, maintains constant body temperature   (7)

b. All living organisms display certain characteristics. Complete the following sentences about these characteristics.

respiration i.   releasing energy in cells is called…   (1)
excretion ii.   the removal of the waste products of the chemical processes in the body is called…   (1)
sensitivity iii.   the ability to respond to changes in the environment is called…   (1)

3. a. Diagrams A to H show organisms or parts of organisms (not drawn to scale).

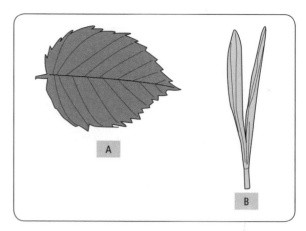

i. State which of the drawings shows a monocotyledon leaf. State **one** reason for your choice.   (1)

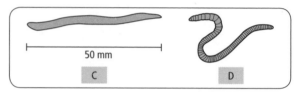

ii. State which of the drawings shows an annelid. State **one** reason for your choice.   (1)

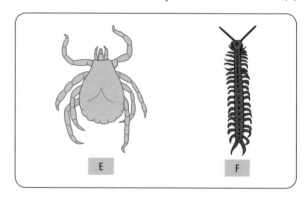

iii. State which of the drawings shows an arachnid. State **one** reason for your choice.   (1)

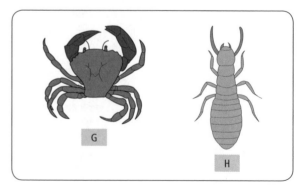

iv. State which of the drawings shows a crustacean. State **one** reason for your choice.   (1)

b. The length of the drawing of worm **C** is shown. The actual length of the worm is 5 mm. Calculate the magnification of this drawing. Show your working.   (2)

CIE 0610 June '07 Paper 2 Q1

4. The diagram below shows a mayfly nymph (a larva) that lives in water.

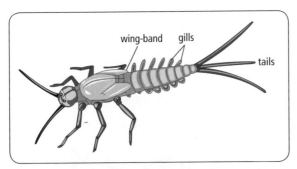

a. i   List two features, visible in the diagram, that show this is an insect.   (2)
    ii. What special adaptation does the insect shown above have that allows it to live in water?   (1)
b. Below there are five mayfly nymphs.

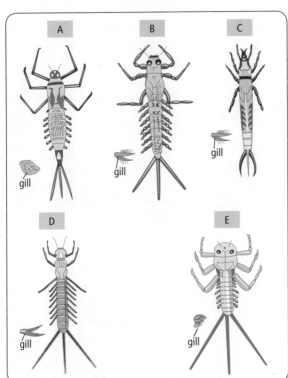

Use the key below to identify the species of each mayfly.

		SPECIES
1	Rear pair of legs point towards tails ---------------------- go to 2 Rear pair of legs point forwards or sideways --------- go to 3	
2	Gills project sideways from body Gills folded over body	*Paraleptophlebia* *Ephemera*
3	Each gill a single flat plate ---------------------------------- go to 4 Each gill divided into two strands	*Potomanthus*
4	Tails 'feather' like in shape Tails 'needle' shaped	*Centroptilum* *Ecdyonurus*

Write the diagram letter of each of the species in the correct box of the table.

SPECIES	DIAGRAM LETTER
Centroptilum	E
Ecdyonurus	A
Ephemera	C
Paraleptophlebia	D
Potomanthus	B

(4)

CIE 0610 June '05 Paper 2 Q1

5. The diagram shows ten organisms, **A** to **J**.

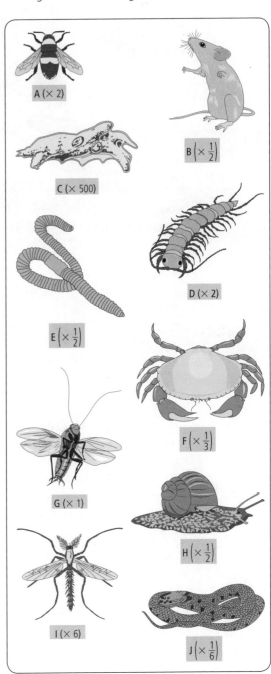

a. Look at the scale factor next to each organism. In real life,
   which organism is the smallest?
   which organism is the largest? (2)
b. **Five** of the organisms in the diagram have several features in common.
   List the **letters** of these five organisms. (2)
c. i. Choose **three** of the organisms from your list in (b) which look to be more closely related to each other than they are to the other two organisms. (1)
   ii. Give **two** features of these organisms which the other two organisms do **not** have. (2)

6. The drawings show four reptiles.

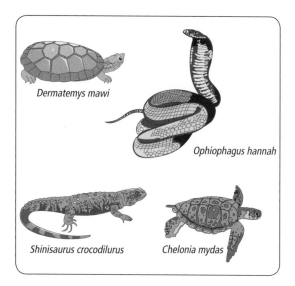

Dermatemys mawi

Ophiophagus hannah

Shinisaurus crocodilurus

Chelonia mydas

a. Complete the following key so that it can be used to identify each of the four reptiles. (4)
   Key:
   1   Has legs   go to 2
       No legs
   2   Most of body covered by a shell
                                 Shinisaurus crocodilurus
   3   Front legs   *Dermatemys mawi*
       Front legs   *Chelonia mydas*
b. Describe **one** feature, **which you can see in the drawings**, which all four reptiles have in common, but which is not present in a human. (1)

7. A large number of seeds were germinated on damp sand. Random samples of 10 seedlings were taken every two days.

The fresh mass and the dry mass of each sample were measured and are shown in the graph below.

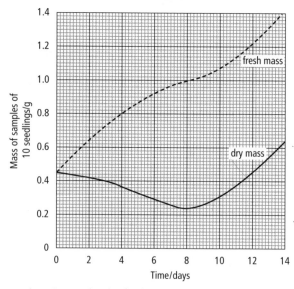

a. i. State why the fresh mass and dry mass of a seedling are different.
   ....
   .... (1)
   ii. Fresh mass is not reliable as a measure of plant growth.
   Suggest why dry mass is a more reliable measure of plant growth.
   ....
   .... (1)
   iii. Explain why 10 seedlings, rather than 1, were used for each sample.
   ....
   ....
   .... (1)
b. i. Describe what happens to the **fresh** mass of the seedlings in the first 2 days after the seeds were set to germinate.
   ....
   .... (2)
   ii. Suggest a reason for this change in mass.
   ....
   .... (1)
c. i. Describe what happens to the **dry** mass of the seedlings during the first 8 days.
   ....
   .... (1)
   ii. Suggest a reason for this change in mass.
   ....
   ....
   .... (2)
d. Suggest which processes begin in the living seed during the early stages of germination.
   ....
   ....
   ....
   ....
   .... (4)

LIVING ORGANISMS AND THEIR CHARACTERISTICS: Crossword

ACROSS:
1. The generation of new individuals of the same species
6. Single celled organism with a definite nucleus
9. The ability to detect changes in the environment
10. A sequence of DNA – may be used in classification
12. This kingdom includes humans and other chordates
13. A change in the form or function of an organism
16. Oxidation of food to release energy
21. Measurable differences between organisms
22. A change in position of an organism
23. The peak of the hierarchy of taxonomic groups
24. Kingdom whose members feed by external digestion
25. Essential to drive the processes of life and to maintain organization

DOWN:
2. Removal of the waste products of metabolism
3. Genus that includes all the big cats
4. The 'proper' name for a human – means a 'wise man'
5. Single cell without a definite nucleus
7. The 'two part' system of naming living organisms
8. The father of naming organisms?
11. The arrangement of molecules and cells into a working pattern
14. Supply of raw materials for metabolism, growth reproduction
15. A permanent increase in size
17. Another name for 'classification' – the science of putting living organisms into groups
18. Kingdom with members capable of photosynthesis
19. Members of this taxonomic group can mate and produce fertile offspring
20. A set of questions that can be used to 'unlock' the name of organism!

19

below show five different organisms.

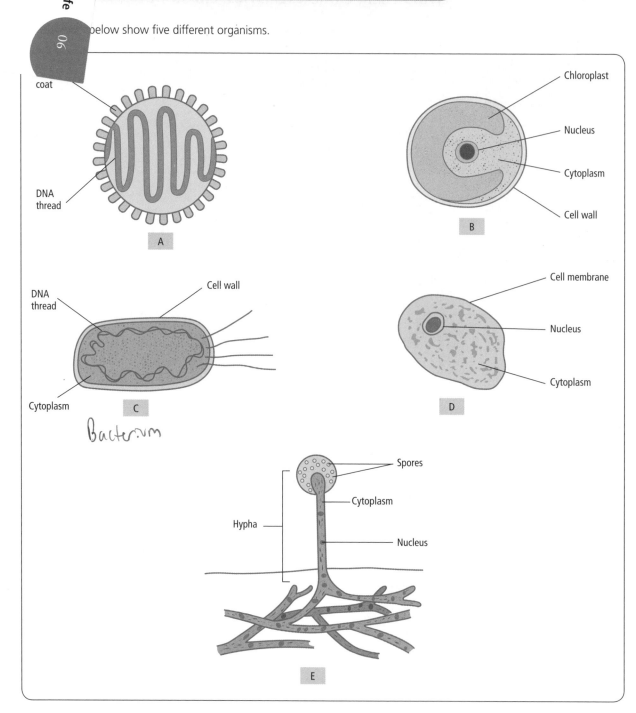

coat

A

DNA
thread

Chloroplast

Nucleus

Cytoplasm

Cell wall

B

DNA
thread

Cell wall

Cytoplasm

C

Bacterium

Cell membrane

Nucleus

Cytoplasm

D

Spores

Cytoplasm

Hypha

Nucleus

E

a. i.   Which of the organisms shown is a
         bacterium?   (1)
    ii. Give **two** reasons for your choice.   (2)
    iii. Which of the organisms belongs to a
         group all of whose members are
         parasitic?   (1)
    iv. What is meant by a parasite?   (2)

b. i.   Which two types of the organisms shown are
         decomposers?   (2)
    ii. What would be the importance of organism
         B in a food chain?   (2)

2. The diagram below shows the feeding process in a small organism.

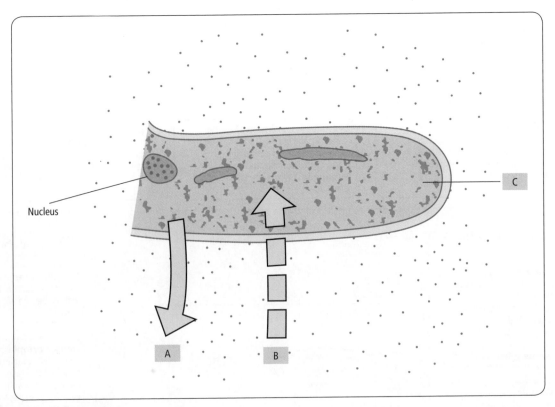

Nucleus

C

A

B

▲ **Fig. 56:** Feeding hypha.

The organism is feeding on some waste protein – the skin of a dead animal.

a. Which **class** of enzyme must be secreted at A? (1)
b. Which type of compound will be absorbed at B? (1)
c. Which Kingdom does this organism belong to? (1)
d. What is the name given to the feeding structure labelled C? (1)

A biologist was interested in the uptake of the compounds at B. He carried out an experiment to investigate the effect of oxygen concentration on this process. The results are shown in the table below.

OXYGEN CONCENTRATION (%)	0	4	8	12	16	20
RATE OF UPTAKE (ARBITRARY UNITS)	2	10	18	25	32	32

e. Draw a graph of this information. (5)
f. What does the graph tell you about the uptake of compounds at B? (3)
g. Suggest another factor that might affect the rate of uptake of these compounds. (1)
h. How could this information be useful to an organic farmer? (2)

3. Indicate which of the following statements is true (T) and which is false (F). (10)

a. Bacteria always cause disease.
b. A typical bacterium is about one thousandth of a metre long.
c. Viruses can only reproduce inside other living cells.
d. Bacteria are smaller than viruses.
e. Viruses do not have their own genes.
f. All viruses are the same shape.
g. Viruses are simpler in structure than bacteria.
h. Viruses always cause disease.
i. Viruses can only be seen with an electron microscope.
j. Bacteria can infect animals and plants.

Cells: building blocks of life

Living things are made of cells. Many of the chemical reactions that keep organisms alive (metabolic functions) take place in cells.

COMMON FEATURES OF CELLS

All cells (with a few exceptions) have these three things:

Cell membrane: a thin 'skin' that surrounds the cell contents. It controls the passage of dissolved substances into and out of the cell. This membrane is selectively permeable.

Cytoplasm: the contents of the cell (except for the nucleus). It is made up of water and dissolved substances. It also contains small structures **(organelles)** where chemical reactions take place.

Nucleus: the 'control centre' of the cell. It contains the genetic material **(DNA)** that carries the instructions that control the structure and activities of the cell. (Red blood cells do not have nuclei.)

PLANT CELL FEATURES

Cell wall: a rigid (stiff) cell wall made of cellulose. This gives support. As a result, plant cells are fairly regular in shape. Water and dissolved substances can pass through the **permeable** cell wall.

Vacuole: the large, permanent vacuole contains water and dissolved substances **(cell sap)**. This helps to maintain pressure in the cells.

Chloroplasts: these contain **chlorophyll** and the **enzymes** needed for photosynthesis. They are found in the cells of green plants.

Stored food (starch): photosynthesis produces glucose (sugar). This is converted into **starch** and stored in the cytoplasm.

ANIMAL CELL FEATURES

Irregular shape: animal cells do not have a rigid cell wall so they are irregular in shape.

Denser cytoplasm: animal cells contain more dissolved substances and more organelles than plant cells. For example, animal cells contain more of the rod-like structures called **mitochondria** where respiration takes place. This is so they can release lots of energy quickly for fast movement.

Stored food (glycogen): carbohydrates are stored as glycogen in animal cells.

Vacuoles: animal cells may have several small, temporary vacuoles. These can be for digestion or the excretion of excess water.

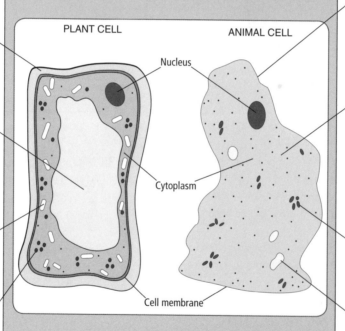

PLANT CELL   ANIMAL CELL

Nucleus

Cytoplasm

Cell membrane

WORKING OUT THE SIZE OF CELLS: you will need to **measure** something, and use a **magnification** to work out an **actual** size

$$\text{MAgnification} = \frac{\text{Measured length}}{\text{Actual length}}$$

so

$$\text{Actual length} = \frac{\text{Measured length}}{\text{Magnification}}$$

Cells, tissues, organs and organ systems

Multicellular plants and animals contain many different types of cell. Each type of cell is designed for a particular function. Cells are organized to form tissues, organs, and organ systems. In a healthy **organism**, all the systems work together.

SPECIALIZED CELLS

A specialized cell is designed to do a particular job.
- Nerve cells have long fibres to carry messages.
- Muscle cells can contract and relax.
- Red blood cells carry oxygen around the body. They contain **haemoglobin**, which can combine with oxygen.

Specialized cells
red cells carry oxygen
white cells attack bacteria
platelets help clotting

TISSUES

Large numbers of specialized cells together make up **tissue**.

Muscle, blood and nerves are all tissues.

Blood tissue contains red cells for carrying oxygen, white cells for destroying harmful bacteria, and **platelets** to cause clotting in cuts.

Tissues
blood vessels (capillaries)
muscles
(especially in the heart)
blood

NB Arteries and veins are usually thought of as organs as they consist of several tissue layers.

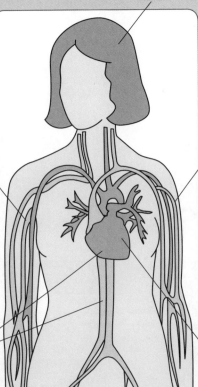

Organism
human

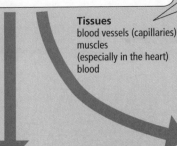

Organ system
blood circulation system

Organ
heart (to pump blood)

ORGANISM

Various organ systems together make up an **organism**.
You are a human organism. You have:
- a respiratory system
- a digestive system
- a circulatory system
- a nervous system
- an endocrine system

ORGAN SYSTEMS

Various organs together make up an **organ system**. For example, the **circulatory system** carries blood to all parts of the body. It is made up of the heart, the arteries, the veins, the capillaries and, of course, the blood.

ORGANS

Various tissues together make up an **organ**. Each organ has its own specific job. The heart, the stomach and the brain are all organs.

The **heart** has to pump blood around the body. It is made up of **muscle tissue**, **blood vessels** and **nerves**.

1. The diagram below shows several types of animal and plant cell (not drawn to scale).

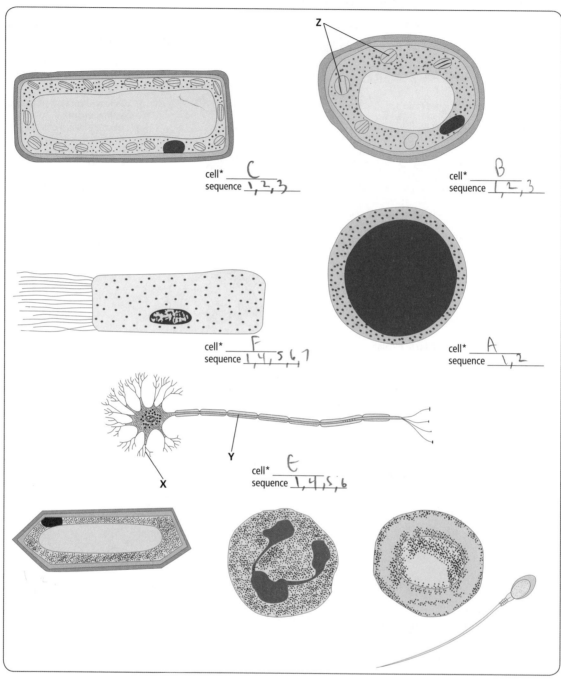

cell* __C__
sequence __1, 2, 3__

cell* __B__
sequence __1, 2, 3__

cell* __F__
sequence __1, 4, 5, 6, 7__

cell* __A__
sequence __1, 2__

cell* __E__
sequence __1, 4, 5, 6__

(5)

a. Use the key below to identify each of the cells marked with an asterisk (*). Write the letter corresponding to each cell next to the appropriate diagram. For each of the asterisked cells write the sequence of numbers from the key that led to your answer.

Identification Key

1   Cell with distinct cell wall .... go to 2
    Cell with membrane but no cell wall .... go to 4
2   Cell with chloroplasts in the cytoplasm .... go to 3
    Cell without chloroplasts in the cytoplasm ... CELL A
3   Cell with less than 10 chloroplasts visible ..... CELL B
    Cell with more than 10 chloroplasts visible .. CELL C

4   Cell containing a nucleus .... go to 5
    Cell lacking a nucleus .... CELL D
5   Cell with projections at one or more ends ... go to 6
    Cell without projections .... go to 8
6   Cell with projections at each end .... CELL E
    Cell with projection/projections at one end only ......
    .... go to 7
7   Cell with a large number of projections at one end .
    .... CELL F
    Cell with a single, long projection at one end ....
    .... CELL G
8   Cell with a many-lobed nucleus .... CELL H
    Cell with a round nucleus .... CELL I

b. **Three** of the cells shown in the diagram are types of human blood cell. Name each of the three cells and describe **ONE** function of each of the cells.
 i. Name of cell
 Function

ii. Name of cell
 Function
iii. Name of cell
 Function   (3)

2. The diagrams below show a plant cell, an animal cell and a virus. The diagrams are NOT drawn to the same scale.

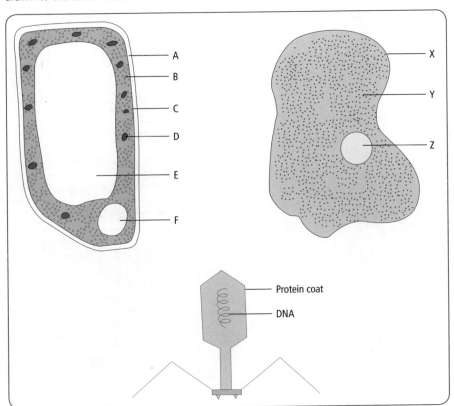

a. The parts of the plant cell are labelled A, B, C, D, E and F. Write the letter of the named part in the box next to its name in the list below.

CELL WALL — A

CYTOPLASM — B

VACUOLE — C   (3)

b. Name the part labelled Z in the diagram of an animal cell.   (1)
c. State TWO differences, shown in the diagrams, between the plant cell and the animal cell.   (2)
d. Which part of the plant cell contains the genes/alleles?   (1)
e. What biological term describes a group of similar cells?   (1)
f. Use ONLY the information in the diagram to suggest TWO reasons why the virus is not a cell.   (2)

3. The diagram shows a cell from a plant leaf.

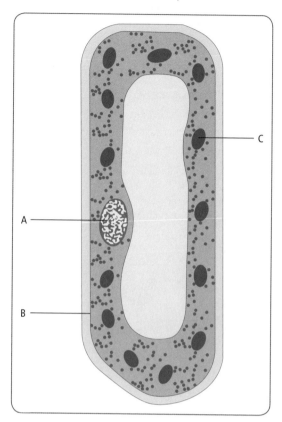

a. Name structures **A** and **B**. (2)
b. Structure **C** is a chloroplast. What is the function of a chloroplast? (1)
c. A plant cell has chloroplasts where as an animal cell does not.
   Give **two** more differences between plant and animal cells. (2)
d. An average plant cell is 50 μm long and 20 μm wide. How many plant cells could fit into 1 mm²? Show your working. (1)

4. a. The diagrams below show five types of tissue. Match each tissue with its correct function. (5)

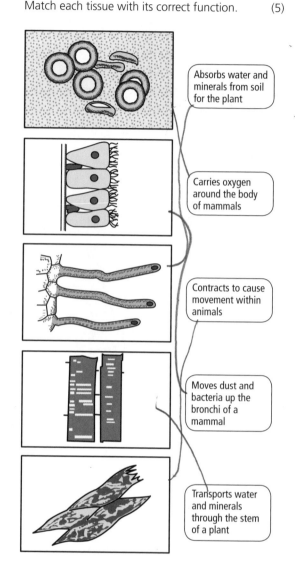

Absorbs water and minerals from soil for the plant

Carries oxygen around the body of mammals

Contracts to cause movement within animals

Moves dust and bacteria up the bronchi of a mammal

Transports water and minerals through the stem of a plant

b. Explain why the heart is described as an organ and not as a tissue. (2)

REVISION SUMMARY: Fill in the missing words

Use words from this list to complete the following paragraphs. The words may be used once, more than once or not at all.

PALISADE CELL, EPIDERMIS, TISSUES, EXCRETORY SYSTEM, SPECIALISED, CELLS, BLOOD, KIDNEY, CHLOROPLASTS, LEAF, RED BLOOD CELL, DIVISION OF LABOUR, XYLEM, PHLOEM, NERVOUS, SYSTEMS, ENDOCRINE, ORGAN

a.   Large numbers of ....cells........that have the same structure and function are grouped together to form .....tissues....., for example .....blood.......... Several separate tissues may be joined together to form an ......organ........., which is a complex structure capable of performing a particular task with great efficiency. In the most highly developed organisms these complex structures may work together in ....systems..., for example the ...excretory...system..in humans is responsible for the removal of the waste products of metabolism.   (6)

b.   The structure of cells may be highly adapted to perform one function, i.e. the cells may become ...specialised... One excellent example is the ...RBC.............. which is highly adapted to carry oxygen in mammalian blood. If the different cells, tissues and organs of a multicellular organism perform different functions they are said to show ....D.O.L........... One consequence of this is the need for close co-ordination between different organs – this function is performed by the ...nervous......... and ......endocrine.. systems in mammals.   (5)

c.   In plants an example of a cell highly specialized for photosynthesis is the .....palisade..cell, which contains many ...chloroplasts... These cells are located in the organ called the ...leaf..............., which also contains other tissues such as ....xylem......, which limits water loss, and .....phloem....., which transports water and mineral ions to the leaf.   (5)

Chapter 9:
Movement of molecules in and out of cells

Diffusion, osmosis and active transport

are processes by which molecules are moved. Diffusion and osmosis are passive, but active transport requires energy.

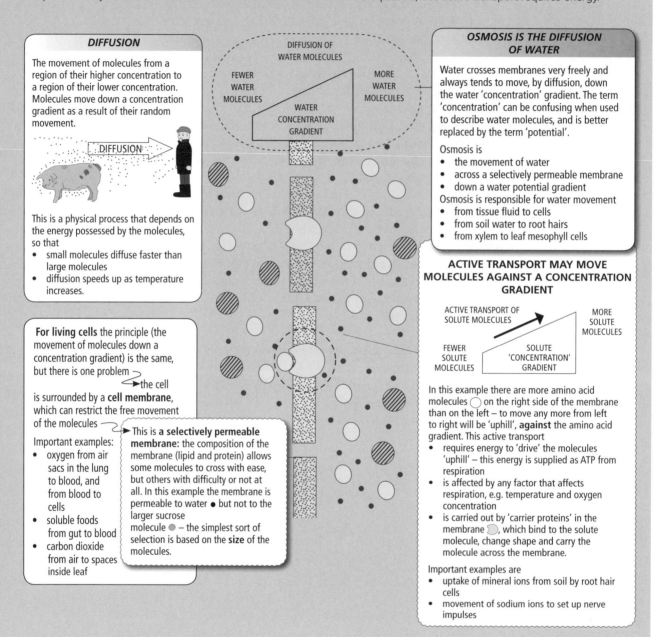

DIFFUSION

The movement of molecules from a region of their higher concentration to a region of their lower concentration. Molecules move down a concentration gradient as a result of their random movement.

DIFFUSION

This is a physical process that depends on the energy possessed by the molecules, so that
- small molecules diffuse faster than large molecules
- diffusion speeds up as temperature increases.

For living cells the principle (the movement of molecules down a concentration gradient) is the same, but there is one problem → the cell is surrounded by a **cell membrane**, which can restrict the free movement of the molecules

Important examples:
- oxygen from air sacs in the lung to blood, and from blood to cells
- soluble foods from gut to blood
- carbon dioxide from air to spaces inside leaf

→ This is **a selectively permeable membrane**: the composition of the membrane (lipid and protein) allows some molecules to cross with ease, but others with difficulty or not at all. In this example the membrane is permeable to water ● but not to the larger sucrose molecule ◒ – the simplest sort of selection is based on the **size** of the molecules.

DIFFUSION OF WATER MOLECULES

FEWER WATER MOLECULES

MORE WATER MOLECULES

WATER CONCENTRATION GRADIENT

OSMOSIS IS THE DIFFUSION OF WATER

Water crosses membranes very freely and always tends to move, by diffusion, down the water 'concentration' gradient. The term 'concentration' can be confusing when used to describe water molecules, and is better replaced by the term 'potential'.

Osmosis is
- the movement of water
- across a selectively permeable membrane
- down a water potential gradient

Osmosis is responsible for water movement
- from tissue fluid to cells
- from soil water to root hairs
- from xylem to leaf mesophyll cells

ACTIVE TRANSPORT MAY MOVE MOLECULES AGAINST A CONCENTRATION GRADIENT

ACTIVE TRANSPORT OF SOLUTE MOLECULES

MORE SOLUTE MOLECULES

FEWER SOLUTE MOLECULES

SOLUTE 'CONCENTRATION' GRADIENT

In this example there are more amino acid molecules ◯ on the right side of the membrane than on the left – to move any more from left to right will be 'uphill', **against** the amino acid gradient. This active transport
- requires energy to 'drive' the molecules 'uphill' – this energy is supplied as ATP from respiration
- is affected by any factor that affects respiration, e.g. temperature and oxygen concentration
- is carried out by 'carrier proteins' in the membrane ◖, which bind to the solute molecule, change shape and carry the molecule across the membrane.

Important examples are
- uptake of mineral ions from soil by root hair cells
- movement of sodium ions to set up nerve impulses

---

1. The cells of the epidermis of the rhubarb stem contain a red pigment. This pigment makes it easy to see the cytoplasm in these cells, and to observe how the cytoplasm and cell membrane are affected by the external environment.

Strips of rhubarb epidermis were placed in three different sucrose solutions. The strips were left in the solutions for 30 minutes, and the cells were observed under the high power objective lens of a light microscope. The observations are recorded in the table opposite.

SOLUTION	APPEARANCE
A	Cell membrane pulled away from cell wall. No visible vacuole.
B	Cell membrane pushed firmly against cell wall. Large vacuole near centre of cell.
C	No change from 'normal' appearance – cell membrane just touching cell wall. Medium-sized vacuole in centre of cell

a. Name the process responsible for these changes in the cells' appearance.

b. What term is used to describe the cell membrane that allows this process to occur?

c. What has happened to cause the cell membrane and cytoplasm to push against the cell wall in solution B?

d. What can you say about solution C to explain the appearance of the cells? (4 × 1)

e. Copy this diagram. Complete it, and add labels, to show the cell membrane, cytoplasm and vacuole of a cell placed in solution A. (3)

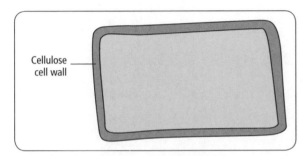

Cellulose cell wall

2. An experiment was set up to try to measure the concentration of the cytoplasm inside potato cells. A cork borer was used to produce identical cylinders of potato tissue. Twenty-one of these cylinders were blotted dry and weighed, and then three were placed into each of seven boiling tubes labelled A to G. The boiling tubes contained a series of sugar solutions, ranging from 0.0 to 0.6 molar.

Two hours later the cylinders were removed from the tubes, blotted dry and reweighed. The results are shown in the table below.

TUBE	A	B	C	D	E	F	G
Concentration of sugar solution / molar	0.0	0.1	0.2	0.3	0.4	0.5	0.6
Percentage change in mass of potato cylinders	+10	+6	+2	−3	−7	−10	−12

a. Plot a graph of percentage change in mass against concentration of sucrose solution. (4)

b. Use your graph to state the concentration of sucrose solution that will give no change in mass of the potato tissue. (2)

c. Use your biological knowledge to explain how this value allows you to work out the concentration of the cytoplasm in the potato cells. (2)

d. Explain why **three** potato cylinders were placed into each of the solutions. (2)

e. Explain why the potato cylinders are blotted dry before they are weighed. (2)

f. Three more potato cylinders were placed into a 1 molar solution of sucrose. The potato cylinders weighed a total of 10 g. Predict the mass of this tissue after two hours in this solution. (2)

3. Crop production in many areas of the world needs the application of large volumes of water. However, when the water evaporates from the soil, traces of salts are left behind. After several years, the soil becomes too salty for most plants to grow in it.

a. i. State three functions of water in plants. (3)

ii. With reference to the water potential gradient, explain why plants may die when grown in salty soil. (3)

b. Some plants are able to pump salts out of their roots.

i. Name the process plants could use to pump salts out of their roots. (1)

ii. Suggest how the process named in **(i)** could affect the rate of growth of the plants if the process was operating all the time. (2)

iii. Plants need mineral salts for normal, healthy growth. Name two minerals that plants need and state their functions. (4)

Adapted from CIE 0610 June '06 Paper 3 Q5

SUMMARY REVIEW: Fill in the missing words

Use words from this list to complete the following paragraphs. You may use each word once, more than once or not at all.

DIFFUSION, OSMOSIS, PHOTOSYNTHESIS, ACTIVE TRANSPORT, CELL WALL, SWELL, SHRINK, PARTIALLY PERMEABLE MEMBRANE, CYTOPLASM, RESPIRATION, OXYGEN, GLYCOGEN, CARBON DIOXIDE, AMINO ACIDS, PATHWAY, ENERGY, ALONG, AGAINST, CONCENTRATION GRADIENT, EPIDERMIS

Animal cells contain ...., a semi-fluid solution of salts and other molecules, and are surrounded by a .... When surrounded by distilled water, the animal cells .... because the cell has a .... water potential than the surrounding water. Plant cells do not have this problem because they are surrounded by a ....

In the gut soluble food substances such as .... cross the gut lining into the capillaries by the process of ...., which is the movement of molecules down a .... When an equilibrium is reached between the gut contents and the blood, glucose may continue to be moved using the process called ...., which consumes .... and can move molecules .... a concentration gradient.

The leaves of green plants obtain the gas ...., which they require for the process of photosynthesis, by the process of .... They also lose the gas oxygen, produced during .... by the same process. (14)

Chapter 10:
Biological molecules and food tests

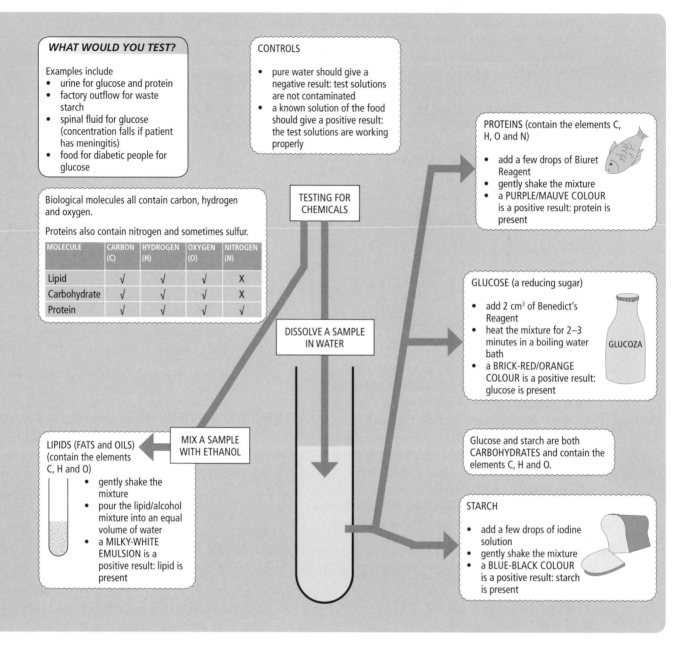

WHAT WOULD YOU TEST?

Examples include
- urine for glucose and protein
- factory outflow for waste starch
- spinal fluid for glucose (concentration falls if patient has meningitis)
- food for diabetic people for glucose

CONTROLS
- pure water should give a negative result: test solutions are not contaminated
- a known solution of the food should give a positive result: the test solutions are working properly

PROTEINS (contain the elements C, H, O and N)
- add a few drops of Biuret Reagent
- gently shake the mixture
- a PURPLE/MAUVE COLOUR is a positive result: protein is present

Biological molecules all contain carbon, hydrogen and oxygen.

Proteins also contain nitrogen and sometimes sulfur.

MOLECULE	CARBON (C)	HYDROGEN (H)	OXYGEN (O)	NITROGEN (N)
Lipid	√	√	√	X
Carbohydrate	√	√	√	X
Protein	√	√	√	√

TESTING FOR CHEMICALS

DISSOLVE A SAMPLE IN WATER

GLUCOSE (a reducing sugar)
- add 2 cm³ of Benedict's Reagent
- heat the mixture for 2–3 minutes in a boiling water bath
- a BRICK-RED/ORANGE COLOUR is a positive result: glucose is present

GLUCOZA

Glucose and starch are both CARBOHYDRATES and contain the elements C, H and O.

MIX A SAMPLE WITH ETHANOL

LIPIDS (FATS and OILS) (contain the elements C, H and O)
- gently shake the mixture
- pour the lipid/alcohol mixture into an equal volume of water
- a MILKY-WHITE EMULSION is a positive result: lipid is present

STARCH
- add a few drops of iodine solution
- gently shake the mixture
- a BLUE-BLACK COLOUR is a positive result: starch is present

1. Food contains biological molecules. It is sometimes important to know exactly which molecules are present in which foods. There are chemical tests called FOOD TESTS that allow us to find this out.

    The boxes opposite list certain molecules and food tests.

    a. The molecules and the chemical tests are only correctly matched in four of the boxes. Write down the letters of these four boxes. (4)

    b. Describe exactly how you would carry out a Benedict's test if you were given a powdered sample of a food. Be sure to include any safety measures you would take. (3)

A	Protein	Biuret
B	Simple sugar	Iodine solution
C	Fat	Biuret
D	Starch	Benedict's reagent
E	Protein	Alcohol emulsion
F	Starch	Iodine solution
G	Fat	Alcohol emulsion
H	Simple sugar	Benedict's reagent

2. The figure below shows how four different food tests
(**1**, **2**, **3** and **4**) were carried out. Each gave a positive
result.

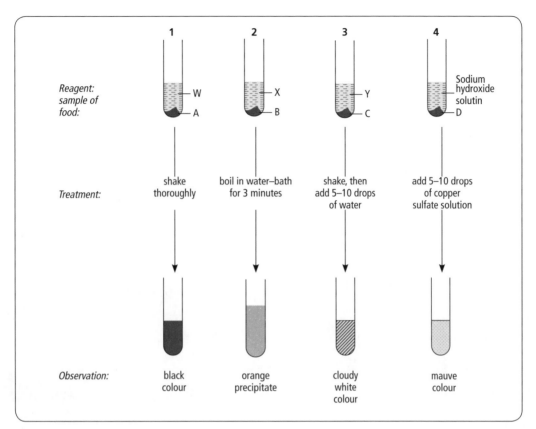

	1	2	3	4
Reagent: sample of food:	W / A	X / B	Y / C	Sodium hydroxide solutin / D
Treatment:	shake thoroughly	boil in water–bath for 3 minutes	shake, then add 5–10 drops of water	add 5–10 drops of copper sulfate solution
Observation:	black colour	orange precipitate	cloudy white colour	mauve colour

a. Complete the table giving the names of the
reagents **W**, **X** and **Y**, and the type of nutrient
shown to be present by each of the four tests.

	TEST			
	1	2	3	4
Reagent				
Nutrient in sample of food				

(5)

b. State the colours you would have seen if the results
were negative in:
   test 1
   test 2
   test 4 (3)
c. i   Suggest **two** possible items of a person's diet
      which are rich in the nutrient in the food **C**.
   ii Suggest **two** possible items of a person's diet
      which are rich in the nutrient in the food **D**.

(2)

3. It is possible to carry out simple chemical tests to
investigate the chemical content of biological solutions.
Usually these tests involve mixing reagents with the
solution, and noting any colour change. The following
are the results obtained from a series of these tests.

	COLOUR AFTER TESTING WITH REAGENT			
SOLUTION	IODINE SOLUTION	BENEDICT'S REAGENT	BIURET REAGENT	BENEDICT'S AFTER ACIDIFICATION AND NEUTRALIZATION
A	Blue-black	Clear blue	Clear blue	Clear blue
B	Straw yellow	Orange	Purple	Orange
C	Straw yellow	Clear blue	Clear blue	Orange
D	Blue-black	Clear blue	Faint purple	Clear blue
E	Straw yellow	Orange	Clear blue	Orange

a. The waste water from a laundry (1)
b. Milk (1)
c. Crushed potato (1)
d. Urine from a sufferer from sugar diabetes (1)
e. Sweetened tea (1)

4. The bar chart below shows the percentage of each of the main elements in the human body.

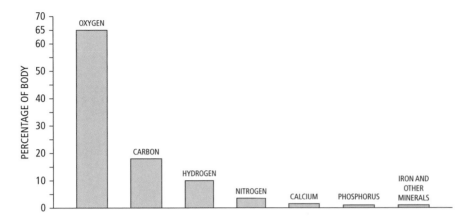

a. Convert this information
   i. into a table
   ii. into a pie chart (5)
b. Which do you think is the best way of showing the data? Explain why. (2)
c. These elements are largely present as molecules. The proportions of the main molecules in a human body are approximately:

   Fats   14%
   Proteins   12%
   Carbohydrates   1.0%
   Water   70%

   i. Which is the most abundant molecule? (1)
   ii. Which molecule contains most of the nitrogen? (1)
   iii. Which molecule contains most of the oxygen? (1)
   iv. Which molecule contains most of the carbon? (1)
   v. The total of these molecules does not add up to 100 per cent. Suggest another organic compound that forms a proportion of the remainder. (1)
   vi. Which structure(s) in the body will contain most of the calcium? (1)

5. Simple chemical tests on foods can show which type of nutrients they contain.

A group of students was given samples of four different foods to test – the samples were prepared by dissolving powdered food in distilled water. The four foods were identified as A, B, C and D. They were also given a sample of distilled water.

They carried out three tests on the five samples. Test 1 was for glucose, test 2 was for protein and test 3 was for starch. The results of the tests are shown in the table.

FOOD TEST	1 GLUCOSE	2 PROTEIN	3 STARCH
Sample A	Blue	Purple	Brown
Sample B	Orange	Blue	Brown
Sample C	Blue	Blue	Black
Sample D	Orange	Purple	Brown
Distilled water	Blue	Blue	Brown

a. which colour indicates the presence of starch? (1)
b. what is the name of the solution which gives a purple colour if protein is present? (1)
c. which one of the samples contained protein and glucose but no starch? (1)
d. why were the tests also carried out on distilled water? (1)
e. This series of tests does not detect the presence of fat. Describe a test for fat, including a positive result for the test. (3)
f. Explain **one** danger to health of eating too many foods which contain large amounts of fat. (1)

TESTING FOR CHEMICALS: Crossword

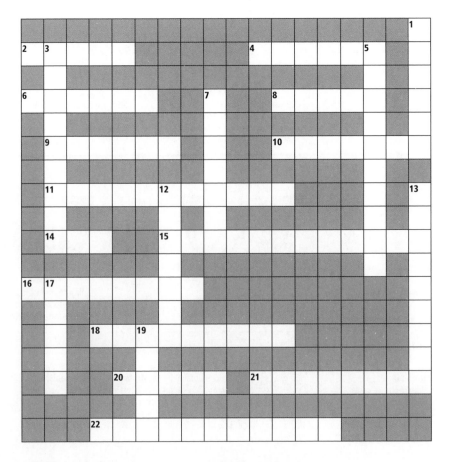

ACROSS:

2 The most common biological molecular–all biochemical reactions take place dissolved in it!
4 Important storage carbohydrate in plants
6 Reagent used to test for protein
8 Biological fluid sampled by doctors testing for anaemia
9 Colour of a positive Benedict's test
10 The most common carbohydrate used for respiration
11 Steroid that can be harmful to blood vessels
14 Abbreviation for the genetic material in plants and animals
15 Reaction in which two molecules are joined together with the elimination of water
16 A test for lipids–millions of tiny fat globules suspended in water
18 Colour of a positive starch test
20 Group of molecules including fats and oils
21 Vital for life, but no single test for them!
22 The most common protein in 8 across

DOWN:

1 Colour of positive test for protein
3 Subunit of a protein (5,4)
5 Reaction in which a large molecule is broken down to smaller ones by the addition of the elements of water
7 Use the Biuret test to check if it's present
12 The sweetest carbohydrate?
13 Reagent that turns orange when heated with 10 across
17 Appearance of the mixture of lipid, ethanol and water
19 The easiest human body fluid to obtain for medical testing

Chapter 11:
Enzymes control biological processes

Enzymes control biological process
and are widely exploited by humans

TEMPERATURE: like all proteins, enzymes are made up of long, precisely folded chains of amino acids. This folding may be 'undone' by high temperature so that the enzyme may lose its active site – it is **denatured**.

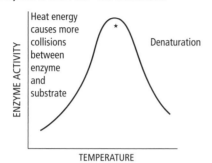

Heat energy causes more collisions between enzyme and substrate

Denaturation

ENZYME ACTIVITY

TEMPERATURE

* The optimum temperature for human enzymes is close to 37 °C. For most plants it is lower.

ENZYMES ARE PROTEINS WHICH ACT AS BIOLOGICAL CATALYSTS

Substrate molecules fit exactly onto an ACTIVE SITE on the enzyme: this is the LOCK AND KEY hypothesis

Substrate molecules react together to form a **product** which leaves the active site

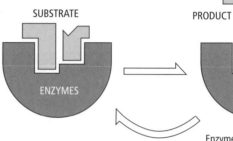

SUBSTRATE

PRODUCT

ACTIVE SITE

ENZYMES

ENZYMES

Enzyme molecule is now free to bind to more substrate molecules

---

pH: is a measure of acidity or alkalinity, and is a mathematical method for expressing the concentration of H^+ ions in solution.

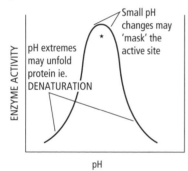

Small pH changes may 'mask' the active site

pH extremes may unfold protein ie. DENATURATION

ENZYME ACTIVITY

pH

* the optimum pH for an enzyme depends on its site of action, pH enzymes in the stomach (where HCl is present) have an optimum about pH 2 but intestinal enzymes (no HCl) have optimum about pH 7.5.

The **shape** of the active site enables the enzyme to 'recognize' its substrate **in a very specific way**. Any factor that alters the enzyme's shape will affect its activity.

Inhibitors and activators: these are molecules that may
- **inhibit** by blocking the active site, e.g. cyanide poisons by blocking enzymes in respiration
- **activate** by helping the active site to achieve its correct shape, e.g. chloride ions in saliva activate the starch digesting enzyme salivary amylase

---

Enzymes may be intracellular or extracellular

these are both made **and** have their action inside cells ('intra' means 'inside')
e.g. photosynthetic enzymes inside chloroplasts or respiratory enzymes inside mitochondria.

these are **made** inside cells but **have their action** outside the cell ('extra' means 'outside')
e.g. digestive systems in the human gut
enzymes released by saprotrophic fungi and bacteria

THE SPECIFICITY AND CATALYTIC ACTIVITY OF ENZYMES MAKES THEM VERY USEFUL TO HUMANS

ENZYMES HELP SEED GERMINATION

- **Amylase** breaks down starch to sugars
- **Lipase** breaks down fats (e.g. in sunflower seeds) to fatty acids and glycerol

GENETIC ENGINEERING

- **Restriction enzymes** are used to cut out specific genes, and to open up bacterial plasmids
- **Ligases** are used to 'stitch' human genes into bacterial plasmids

COMMERCIAL

- **Proteases** help to soften leather for the garment industry
- **Lipase** removes stains from clothing – component of 'biological' washing powders
- **Amylase** converts starch to sugars used in production of syrups, e.g. in fruit pies
- **Pectinase** breaks down small pieces of plant tissue to turn cloudy fruit juice into clear fruit juice

1. Use words from this list to complete the following paragraph about enzymes. You may use each word once, more than once or not at all.

PATHWAYS, ENZYMES, CATALYSTS, ACTIVATORS, PROTEINS, UNUSUAL, SPECIFIC, DENATURED, DESTROYED

Enzymes are .... that speed up the biochemical .... in living organisms. The enzymes themselves are not changed in these reactions, that is they are biological ....
Enzymes are ....– each of them controls only one type of reaction. They are ....by high temperatures and by extremes of pH. (5)

2. In an experiment, apparatus is used to measure the effect of one factor, the **input** or **independent** variable, on the value of a second factor, the **outcome** or **dependent** variable. To be sure that the experiment is a fair test, all other factors must be kept constant – these are **fixed** variables. For example, a simple closed manometer may be used to measure the effect of temperature on the activity of the enzyme catalase.

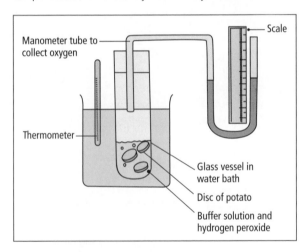

a. What is the input (independent) variable in this experiment?
b. Suggest two fixed variables, and explain how the experimenter could keep them constant.
c. What is the outcome variable in this experiment? How could it be measured?
d. Suggest a suitable control for this experiment.
e. Students using this apparatus collected five sets of data, and used them to calculate a **mean**. How does this improve the **validity** of the results? (7 × 1)

3. Respiration is the process that releases energy in cells. It is a process that depends on enzymes. A group of students were interested in the effects of temperature on the rate of oxygen consumption by maggots. They obtained the following results.

TEMPERATURE / °C	OXYGEN CONSUMPTION / mm per s
15	0.3
25	0.6
35	1.1
45	0.8
55	0.2
65	0.0

a. Plot these results in the form of a graph, and explain the shape of the curve. (6)
b. In this investigation, identify the **input** variable and the **outcome** variable. Suggest any **fixed** variables. (4)

4. a. Define the term *enzyme*. (2)
   b. Enzymes are used in biological washing powders.
      i. Describe how the presence of these enzymes may increase the efficiency of the washing powder in removing stains from clothes. (3)
      ii. Explain why the temperature of the wash needs to be carefully controlled. (3)
      iii. Suggest a suitable temperature for a wash using a biological washing powder. Explain your answer.
         Suitable temperature
         Explanation (1)
   c. Outline how enzymes can be manufactured for use in biological washing powders. (4)

CIE 0610 November '05 Paper 3 Q6

5. Catalase, an enzyme, is present in all living cells including those of potato and liver. It speeds up the breakdown of hydrogen peroxide as shown by the equation:

$$\text{hydrogen peroxide} \xrightarrow{\text{catalase}} \text{oxygen} + \text{water}$$

The oxygen is given off as a gas which can be collected over water, as shown below.

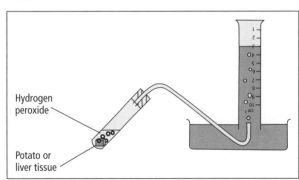

Two different tissues, potato and liver, were used for this investigation. Samples, each of one gram, were prepared from both tissues. Some of the samples were left raw and others were boiled. Some samples were left as one cube and others were chopped into small pieces as shown in the first table below.

2cm³ hydrogen peroxide was added to each sample. The volume of oxygen produced in five minutes was collected in the measuring cylinders, as shown in the table.

SAMPLE	A	B	C	D
Treatment	raw	raw	boiled	boiled
Results for potato				
Results for liver				

a. i. Complete the following table by reading the values for oxygen collected in the measuring cylinders in the table above

TISSUE	VOLUME OF OXYGEN COLLECTED FROM EACH SAMPLE / cm³			
	A	B	C	D
Potato				
Liver				

(2)

ii. Plot the volumes of oxygen collected from the samples as a bar chart.

iii. Describe the difference in results between sample **A** for potato and sample **A** for liver. (2)

iv. There is a difference between the samples for **A** and **B** for liver. Suggest an explanation for this difference. (2)

b. State the importance of samples **C** and **D** in this investigation. (1)

c. Suggest how you could test that the gas given off was oxygen. (1)

Adapted from CIE 0610 June '05 Paper 6 Q1

6. The diagram shows the main features of a biosensor used to monitor blood glucose levels.

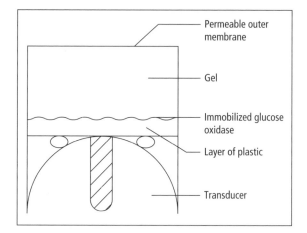

Labels: Permeable outer membrane; Gel; Immobilized glucose oxidase; Layer of plastic; Transducer

a. The biosensor uses the enzyme, glucose oxidase, which is immobilized. Explain what is meant by 'immobilized'. (1)

b. Glucose oxidase catalyses the reaction:

Glucose + Oxygen ⟶ Gluconic acid + Hydrogen peroxide

Glucose diffuses into the biosensor through the outer membrane and gel. Explain how the oxygen concentration around the transducer will be affected. (2)

c. What is the function of the transducer? (1)

ENZYMES CONTROL BIOLOGICAL PROCESSES: Crossword

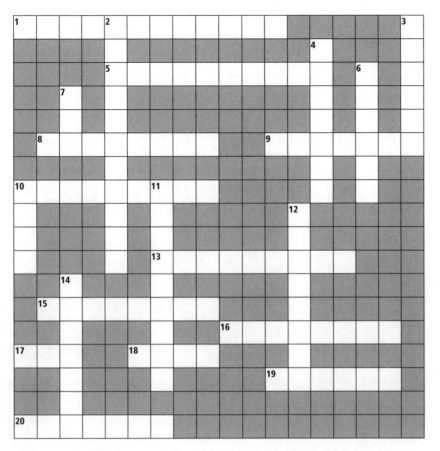

ACROSS:
1   The loss of an enzyme's three dimensional shape
5   All of the chemical reactions occurring in cells
8   More of these ions lower the pH
9   A poison that affects an enzyme in respiration
10 A reactant in an enzyme-catalysed reaction
13 A compound that reduces the activity of an enzyme
15 A protein-digesting enzyme
16 A molecule that speeds up a reaction but is not affected itself
17 With 18 across and 7 down – a mechanism to explain enzyme action
18 see 17 across
19 A stomach enzyme with a very low optimum pH
20 The 'best' value for a factor affecting enzyme activity

DOWN:
2   This factor affects an enzyme's activity – too high and 1 across may be the result
3   Fat-digesting enzyme – useful in biological washing powders
4   Starch-digesting enzyme
6   With 10 down – the part of an enzyme where the reaction occurs
7   See 17 across
10 See 6 down
11 A molecule that can speed up an enzyme-catalysed reaction
12 Enzyme that breaks down harmful ions in liver cells
14 Molecule produced in an enzyme-catalysed reaction

An ideal human diet

contains fat, protein, carbohydrate, vitamins, minerals, water and fibre **in the correct proportions**. For a list of the component atoms of each type of molecule, see Chapter 10 (page 30).

An adequate diet provides sufficient **energy** for the performance of metabolic work, although the 'energy' could be in any form.

A balanced diet provides all dietary requirements **in the correct proportions**. Ideally this would be $\frac{1}{7}$ **fat**, $\frac{1}{7}$ **protein** and $\frac{5}{7}$ **carbohydrate**.

In conditions of **under**nutrition the first concern is usually provision of an **adequate diet**, but to avoid symptoms of **mal**nutrition a **balanced diet** must be provided.

CARBOHYDRATES

Principally as an energy source. **Respiratory substrate** oxidized to release **energy** for active transport, synthesis of macromolecules, cell division and muscle contraction.

Common sources: rice, potatoes, wheat and other cereal grains, i.e. as **starch**, and as refined sugar, **sucrose**, in food sweetenings and preservatives.

Digested in mouth and small intestine and absorbed as **glucose**.

PROTEINS

Are **building blocks** for growth and repair of many body tissues (e.g. myosin in muscle, collagen in connective tissues), as **enzymes**, as **transport systems** (e.g. haemoglobin), as **hormones** (e.g. insulin) and as **antibodies**.

Common source: meat, fish, eggs and legumes/pulses. Must contain eight **essential amino acids** since humans are not able to synthesize them. (Animal sources generally contain more of the essential amino acids than vegetable sources.) Digested in stomach and absorbed as **amino acids**.

Deficiency of protein causes poor growth – in extreme cases (in developing countries) may cause **marasmus** or **kwashiorkor**.

LIPIDS

An energy source (they are highly reduced and therefore can be oxidized to release **energy**). Also important in **cell membranes** and as a component of **steroid hormones**.

Common sources: meat and animal foods are rich in **saturated fats** and **cholesterol**, plant sources such as sunflower and soya are rich in **unsaturated fats**.

Digested in the small intestine and absorbed as **fatty acids and glycerol**.

WATER

WATER is required as a solvent, a transport medium, a substrate in digestive reactions and for lubrication (e.g. in tears). A human requires 2–3 dm³ of water daily – most commonly from drinks and liquid foods.

MINERALS

MINERALS have a range of **specific** roles, and absence may cause **deficiency diseases**
Deficiency causes

	Source	Deficiency causes
e.g. **Calcium** (Ca²⁺)	Dairy products	Poor growth of bones and teeth
Iron (Fe²⁺)	Red meat, spinach	Anaemia – poor oxygen transport in blood

They are usually ingested with other foods, but supplements may be necessary (e.g. iron tablets are sometimes needed following menstruation).

VITAMINS

VITAMINS have no common structure or function, but are essential in small amounts to use other dietary components efficiently. Their absence results in **deficiency diseases**.

	Source	Deficiency causes
e.g. **vitamin C** (water soluble)	Citrus fruits	**Scurvy** – bleeding gums, slow healing of wounds
vitamin D (fat soluble)	Cod liver oil, margarine	**Rickets** – misshapen and poorly growing bones

FIBRE

FIBRE (originally known as **roughage**) is mainly cellulose from plant cell walls and is common in fresh vegetables and cereals. It **may** provide some energy but mainly serves to aid faeces formation and prevent constipation.

DANGER!

Some foods contain additives e.g. PRESERVATIVES (such as nitrates) slow down spoilage of food and COLOURINGS make food attractive. Excess consumption of additives can be dangerous, for example some food colourings can cause **hyperactivity**.

FOOD SUPPLY AND FAMINE

DAMAGE BY PESTS
- especially in monocultures
- pesticides may be overused so resistant pests evolve

ABSENCE OF WATER
- global warming may affect rain fall patterns
- deforestation can affect the water cycle
- diversion of water supplies eg for hydroelectric power schemes

DESERTIFICATION
- soil erosion
- deforestation
- over grazing

FLOODING
- fertile soil can be eroded
- plant roots can be deprived of oxygen
- crop plants can be damaged

COST OF FUEL AND FERTILISER
- machinery may be too expensive to run
- yield may be very low if fertiliser cannot be used

WAR
- crops may be burned to deprive people of food
- farm workers may be forced to join armies
- working on farms may become too dangerous

COST OF TRANSPORTATION
- food surpluses cannot be equally distributed

URBANIZATION
- buildings take up growing space
- cities can affect local temperatures

INCREASING POPULATION
- may be too many people for available food
- domestic animals may graze on land previously used for crops
- next year's seeds may be eaten as food supplies run out

1. The table shows the composition and energy content of four common foods.

FOOD	ENERGY CONTENT / kJ g⁻¹	COMPOSITION PER 100g					
		Protein / g	Fat / g	Carbohydrate / g	Vitamin C / mg	Vitamin D / mg	Iron / mg
A	3700	0.5	80	0	0	40	0
B	150	1.2	0.6	7	200	0	0
C	400	2.0	0.2	25	10	0	8
D	1200	9.0	1.5	60	0	0	0

a. Which food would be best to prevent rickets?
b. Which food would be best for a young man training for cross-country running?
c. Which food would be most needed by a menstruating woman?
d. Which food would be the most useful to a body-builder?
e. Which food would be most dangerous for a person with heart disease? (5)

2. Orange juice is a good source of vitamin C. Three different brands of orange juice were bought from a health store, and another sample was prepared by squeezing fresh oranges. Each of the four 'juices' was tested as shown in the diagram below. A solution containing 0.1 per cent ascorbic acid (vitamin C) was tested at the same time.

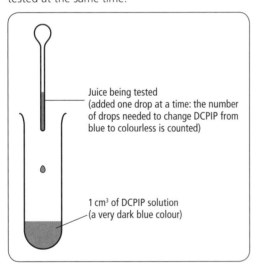

Juice being tested
(added one drop at a time: the number of drops needed to change DCPIP from blue to colourless is counted)

1 cm³ of DCPIP solution
(a very dark blue colour)

The results of this test are shown in the table below.

JUICE TESTED	VOLUME NEEDED TO DECOLOURIZE DCPIP / DROPS
Brand A (long-life)	20
Brand B	13
Brand C	6
Freshly squeezed	5
0.1% ascorbic acid	10

a. Use the following formula to calculate the vitamin C in Brand A.

% vitamin C in Brand A =

$$\frac{(\text{number of drops of 0.1\% ascorbic acid}) \times 0.1}{\text{number of drops of Brand A}} = \ldots\ldots\%$$
(2)

b. Which brand of 'shop' orange juice had the highest vitamin C content? (1)
c. Long-life orange juice is heated to a high temperature, and then sealed in cartons. What effect does this have on the vitamin C content? (2)
d. Give one important function of vitamin C in humans. (1)
e. Name one other source of vitamin C apart from fruit juice. (1)

3. The table shows the amounts of four nutrients required by four people for a balanced diet.

PERSON	PROTEIN / g	IRON / mg	CALCIUM / mg	VITAMIN C / mg
14 year-old boy	66	11	700	25
14 year-old girl	55	13	700	25
30 year-old woman	53	12	500	30
30 year-old pregnant woman	60	14	1200	60

a. i. Explain why there is a difference in the amount of protein required by the 14 year-old boy and the 30 year-old woman. (3)
ii. Explain why there is a difference in the amount of iron required by the 14 year-old girl and the 14 year-old boy. (2)
iii. Explain why there is a difference in the amount of calcium required by the two 30 year-old women. (2)
b. State the role of vitamin C in the human body. (1)

CIE 0610 June '06 Paper 2 Q6

4. A bioreactor may be set up in such a way that a fungus, *Fusarium graminearum*, can be grown on waste materials from the flour and paper industries. The fungus is harvested and processed to make fibres that are dried and pressed to produce **mycoprotein**. Mycoprotein may be flavoured and sold under the trade name Quorn.

The table below provides some nutritional information about a variety of foods.

CONTENT PER 100 g	QUORN	STEAK	CHICKEN	OLIVE OIL	POTATO
Protein / g	12.0	31	24.8	0	4.1
Fat / g	3.2	10.5	5.4	220	0.3
Carbohydrate / g	1.0	0.5	0.2	0	32.4
Cholesterol / g	0	80	74	0	0
Dietary fibre / g	5.1	0	0	0	2.8
Energy / kJ	355	950	620	3600	575

a. Use information from the table to explain why Quorn is regarded as a healthy and nutritious food. (4)

b. Jack produced a meal by frying 250 g of Quorn in 10 g of olive oil, and adding a baked potato. Calculate the total energy value of the meal. Show all of your working. (4)

5. Simple chemical tests on foods can show which type of nutrients they contain.

A group of students was given samples of four different foods to test – the samples were prepared by dissolving powdered food in distilled water. The four foods were identified as A, B, C and D. They were also given a sample of distilled water.

They carried out three tests on the five samples. Test 1 was for glucose, test 2 was for protein and test 3 was for starch. The results of the tests are shown in the table.

FOOD TEST	1: GLUCOSE	2: PROTEIN	3: STARCH
Sample A	Blue	Purple	Brown
Sample B	Orange	Blue	Brown
Sample C	Blue	Blue	Black
Sample D	Orange	Purple	Brown
Distilled water	Blue	Blue	Brown

a. Which colour indicates the presence of starch? (1)
b. What is the name of the solution that gives a purple colour if protein is present? (1)
c. Which one of the samples contained protein and glucose but no starch? (1)
d. Why were the tests also carried out on distilled water? (1)
e. This series of tests does not detect the presence of fat. Describe a test for fat, including a positive result for the test. (3)
f. Explain **one** danger to health of eating too many foods that contain large amounts of fat. (1)

THE HUMAN DIET: Crossword

ACROSS:
1 Deficiency disease that results from a lack of vitamin C in the diet
3 Tasty – but too much can raise blood pressure
4 Mineral required for healthy teeth and bones
7 Should form the bulk of a balanced diet
9 A type of lipid used to produce some hormones
10 The body-building food
11 Essential for insulation and as an energy store
13 Process that releases energy from foods
14 This type of diet has all the essential foods in the correct proportions
15 Mineral component of haemoglobin – deficiency causes anaemia
16 11 across is needed to make this cell boundary

Down
2 Deficiency disease due to low levels of vitamin D
3 Sweet carbohydrate for tea and coffee drinkers
5 Lipid that can cause damage to blood vessels
6 High in protein and dietary fibre – an ideal meal when eaten on toast! (5,5)
8 The result of a poorly balanced diet
12 Essential for the correct formation of faeces

Chapter 13:
Digestion and absorption

Human digestive system

Digestion is the breaking down of large insoluble food molecules into small soluble molecules using mechanical and chemical processes so that they can be absorbed into the bloodstream.

THE DIGESTIVE SYSTEM

Food is digested in the **alimentary canal**. This is a long tube that starts at the mouth, runs through the stomach and intestines and finishes at the anus.

Food is broken down with the help of digestive juices, which contain special chemicals called **enzymes**.

DEALING WITH FOOD

There are five stages in the way we deal with food:

1 **Ingestion** – taking food into the body through the mouth ('eating')

2 **Digestion** – the breakdown of large, insoluble food molecules into small, water-soluble molecules using mechanical and chemical processes

3 **Absorption** – moving digested molecules from the alimentary canal into the bloodstream or the lymph so they can be transported around the body

4 **Assimilation** – movement of absorbed food molecules into cells where they are used and become part of the cells

5 **Egestion** – getting rid of food that could not be digested (e.g. dietary fibre) by passing it as faeces

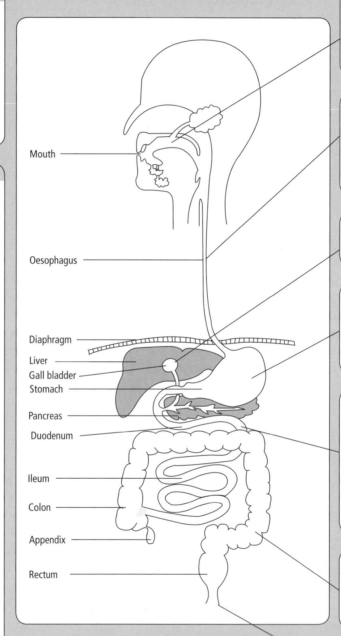

Mouth

Oesophagus

Diaphragm

Liver

Gall bladder

Stomach

Pancreas

Duodenum

Ileum

Colon

Appendix

Rectum

Mouth – Digestion starts here! The teeth cut and grind the food, which is mixed with saliva. The enzyme salivary amylase breaks starch down into maltose (sugar).

Oesophagus – lumps of moist, chewed up food are carried to the stomach by muscle movements. This is called **peristalsis** and also moves partly-digested food along the small intestine.

Gall bladder – stores bile used to help in the digestion of fats.

Stomach – the stomach is like a sack. Here the enzyme **pepsin** breaks big proteins down into small proteins (**polypeptides**). This can take several hours.

Small intestine – this is made up of the **duodenum** and the **ileum**. Here more digestive juices are added. Starches, fats, proteins and complex sugars are broken down into small soluble molecules.

Fully digested food is absorbed into the bloodstream.

Large intestine – this is made up of the **colon** and the **rectum**. Only undigested food reaches here. Water absorbed.

Anus – undigested solid food is passed out as **faeces**.

Digestion, absorption and assimilation

Large food molecules are broken down in our digestive system mostly by chemical reactions. The products are **absorbed** into the blood or lymph then assimilated into cells.

ENZYMES AT WORK

Enzymes are **catalysts**. They speed up the reactions that break down large food molecules, e.g. carbohydrates, proteins and fats.

Enzyme molecules have special shapes. Reactions occur more easily when a food molecule 'locks on' to the **active site of the enzyme**.

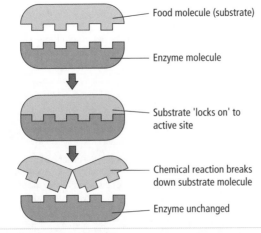

Food molecule (substrate)

Enzyme molecule

Substrate 'locks on' to active site

Chemical reaction breaks down substrate molecule

Enzyme unchanged

The enzymes in our bodies work best at about 37 °C. Some, like those found in the stomach, work best in acid conditions (pH 2). Those found in the duodenum need an alkaline environment (pH 8).

ENZYME (WHERE FOUND)	ACTION
pepsin (stomach)	proteins ⟶ polypeptides (small proteins)
amylase (duodenum)	carbohydrates ⟶ maltose (sugar)
lipase (small intestine)	fats and oils ⟶ fatty acids and glycerol

ABSORPTION IN THE SMALL INTESTINE

The wall of the small intestine has millions of tiny finger-shaped structures called **villi**. These give a huge surface area for absorbing digested food easily.

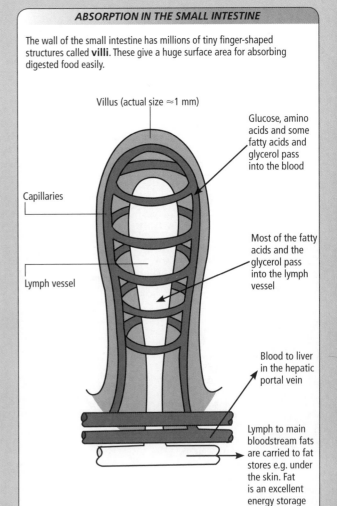

Villus (actual size ≈1 mm)

Capillaries

Lymph vessel

Glucose, amino acids and some fatty acids and glycerol pass into the blood

Most of the fatty acids and the glycerol pass into the lymph vessel

Blood to liver in the hepatic portal vein

Lymph to main bloodstream fats are carried to fat stores e.g. under the skin. Fat is an excellent energy storage substance.

DIGESTED FOOD AND THE LIVER

The liver is the largest organ in the human body. It is like a 'chemical factory'. It processes digested food and other substances in the blood. Its main functions are:
- storing vitamins, minerals and glycogen
- producing special chemicals, e.g. bile for digestion
- processing unwanted substances from the blood, e.g. alcohol is removed from the blood by the liver
- generating heat to keep the body's internal temperature at 37 °C

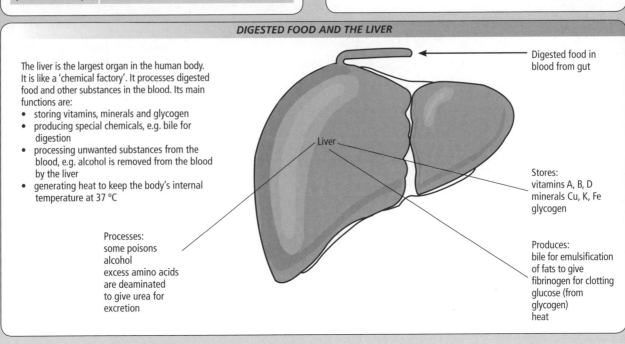

Digested food in blood from gut

Liver

Stores: vitamins A, B, D minerals Cu, K, Fe glycogen

Processes: some poisons alcohol excess amino acids are deaminated to give urea for excretion

Produces: bile for emulsification of fats to give fibrinogen for clotting glucose (from glycogen) heat

TEETH AND TOOTH DECAY

The teeth in the skull

Upper jaw

Incisors– for cutting and biting

Lower jaw

Canine – for holding and cutting

Premolars – for chewing and crushing

Molars (third one is hidden) – for chewing and crushing

CHEWING (MASTICATION) AND SWALLOWING
- the first stage in preparing food for the alimentary canal
- food is cut up by the *teeth* and mixed with *saliva* by the tongue
- the ball of food (the *bolus*) is pushed to the back of the mouth
- swallowing - forcing the bolus into the oesophagus - is a *voluntary action*
- at the same time a *reflex action* lifts the *epiglottis* (a flap of tissue) across the entrance to the larynx, so food does not enter the breathing passages.

A MOLAR TOOTH

Enamel – the hardest tissue in the body. Produced by **tooth-forming cells** and made of calcium salts. Once formed, enamel cannot be renewed or extended.

Cement – similar in composition to dentine, but without any canals. It helps anchor the tooth to the jaw.

Crown

Root embedded in jawbone

Pulp cavity contains:
- **tooth-producing cells**
- **blood vessels**
- **nerve endings** which can detect pain.

Dentine – forms the major part of the tooth. Harder than bone and made of calcium salts deposited on a framework of **collagen fibres**. The dentine contains a series of fine canals which extend to the pulp cavity.

Gum – usually covers the junction between enamel and cement. The gums recede with age.

CARE OF THE TEETH
- reduce number of sweet or acidic foods - fizzy drinks are especially harmful!
- brush to remove sticky food remains that may begin the build-up of *plaque*
- use dental floss to remove material from between teeth

PLAQUE CAUSES TOOTH DECAY

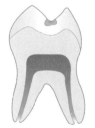

Decay begins in enamel – **no pain.**

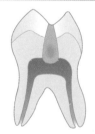

Decay penetrates dentine and reaches pulp – **severe toothache.**

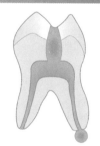

Bacteria now infect pulp and may form abscess at base of tooth – **excruciating pain.**

Fluoride in toothpaste or added to teeth, but
- if added to drinking water we can't control how much we take in
- some teeth develop brown spots when contacted by fluoride - this can be unsightly and reduce confidence.

1. The diagram below shows the human alimentary canal.

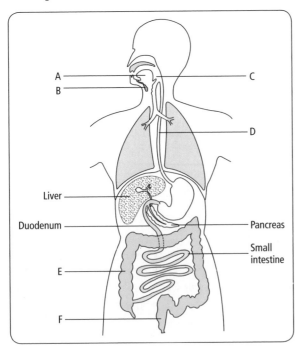

a. Name the part labelled A. (1)
b. Explain how A and B work together to make swallowing more efficient. (2)
c. Why is C important when swallowing? (1)
d. Draw a labelled diagram to explain how food is moved along the structure labelled D. (2)
e. Describe **two** features of the small intestine that help to increase its surface area. (2)
f. The liver has many functions, including the storage of excess carbohydrate.
    i.   Name the carbohydrate used to store excess sugar in the liver. (1)
    ii. Name the blood vessel that carries sugar from the intestine to the liver. (1)
    iii. Name the blood vessel that carries glucose released from the liver some time after feeding. (1)
    iv. The liver also produces bile. What is the function of bile? Where does bile have its effect? (2)
g. The pancreas produces enzymes and releases them into the duodenum.
    i.   How do the enzymes produced in the pancreas get into the duodenum? (1)
    ii. Name one enzyme produced by the pancreas. State what kind of foodstuff it works on, and what the products are. (3)
h. Give one function of each of the parts labelled E and F. (2)

2. Lengths of Visking tubing were set up as shown in the diagram.

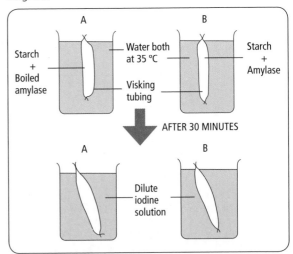

Tube A contained starch and boiled amylase solution. Tube B contained starch and amylase.

Both tubes were left in a water bath at 35 °C for 30 minutes. The tubes were then removed from the water baths, gently blotted with paper towels, and then placed in beakers of dilute iodine solution for five minutes. Visking tubing is permeable to iodine solution.

a. What colour would you expect to see inside A and inside B? (2)
b. Explain why the results are different for tubings A and B. (2)
c. Describe a test you could use to prove your explanation. (3)
d. The drawing below shows a section through some villi.

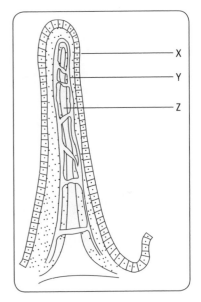

i.   Identify the parts labelled X, Y and Z. (3)
ii. Where in the alimentary canal would villi be found? (1)

iii. Which of the parts X, Y or Z is represented by the Visking tubing? (1)

iv. What is the function of the part labelled Z? (1)

3. The diagram below shows the teeth in the lower jaw of an adult human.

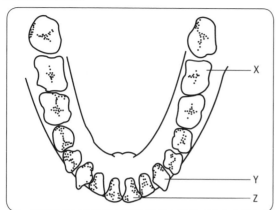

a. i. Name the teeth labelled x, y and z. (3)

ii. Describe the functions of teeth x and z. (2)

b. Name **one** mineral and **one** vitamin that are essential for the healthy development of teeth. (2)

c. The diagram below shows a section through a tooth.

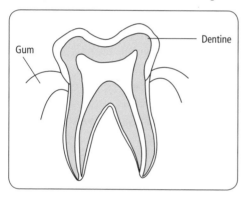

i. Tooth decay is caused by bacteria getting into the dentine. Explain how bacteria can enter the dentine. (3)

ii. List three actions you could take to reduce the risk of tooth decay. (3)

CIE 0610 November '06 Paper 2 Q6

4. The diagram shows the apparatus used to investigate the digestion of milk fat by an enzyme. The reaction mixture contained milk, sodium carbonate solution (an alkali) and the enzyme. In Experiment 1, bile was also added. In Experiment 2, an equal volume of water replaced the bile. In each experiment, the pH was recorded at 2-minute intervals.

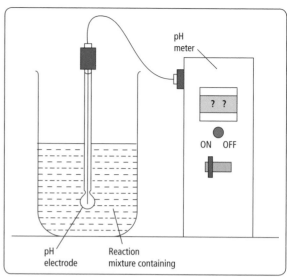

Either: Experiment 1	or: Experiment 2
milk (contains fat)	milk (contains fat)
sodium carbonate solution	sodium carbonate solution
bile	water
enzyme	enzyme

The results of the two experiments are given in the table.

TIME / MINUTES	pH	
	EXPERIMENT 1: WITH BILE	EXPERIMENT 2: NO BILE
0	9.0	9.0
2	8.8	9.0
4	8.7	9.0
6	8.1	8.8
8	7.7	8.6
10	7.6	8.2

a. Milk fat is a type of lipid. Give the name of an enzyme which catalyses the breakdown of lipids. (1)

b. What was produced in each experiment to cause the fall in pH? (1)

c. i. For Experiment 1, calculate the average rate of fall in pH per minute, between 4 minutes and 8 minutes. Show clearly how you work out your final answer. (2)

ii. Why was the fall in pH faster when bile was present? (1)

5. a. i. What name is given to the type of enzyme that breaks down protein in the diet? (1)

ii. What is the product formed when protein is broken down in this way? (1)

b. This table shows the effect of pH on the activity of this type of enzyme.

pH	1.0	2.0	3.0	4.0	5.0	6.0
RATE OF FORMATION OF PRODUCT / MMOLES PER MINUTE	9.5	22.8	16.4	8.1	2.5	0.0

Plot these results in a suitable graph. Supply a title for the graph. (5)

c. Suggest in which part of the human digestive system this enzyme would be working. (1)

d. Why is it necessary to break down large molecules such as protein in the digestive system? (2)

REVISION SUMMARY: Match the term to the definition

	TERM		DESCRIPTION
A	Absorption	1	The conversion of fat globules into smaller droplets
B	Amylase	2	The region of the intestine where most digested food is absorbed
C	Anal sphincter	3	Link between the gall bladder and the small intestine
D	Assimilation	4	Faeces are stored here before they are expelled
E	Bile duct	5	Controls the amount of food leaving the stomach
F	Colon	6	Secretes several enzymes into the small intestine
G	Digestion	7	Muscular contractions that move a bolus of food along the gut
H	Emulsification	8	A protein-digesting enzyme
I	Faeces	9	The organ that produces bile
J	Gall bladder	10	Structures that increase the surface area of the ileum
K	Ileum	11	Region of the intestine in which protein digestion begins
L	Lipase	12	The transfer of digested food from the gut to the bloodstream
M	Liver	13	Region from which most water is absorbed from the contents of the intestine
N	Mastication	14	A fat-digesting enzyme
O	Pancreas	15	A mixture of water, enzymes and mucus
P	Pepsin	16	The chewing and mixing of food with saliva
Q	Peristalsis	17	The use of food molecules in the body
R	Pyloric sphincter	18	The site for bile storage
S	Rectum	19	The breakdown of food into small, soluble molecules ready for absorption
T	Saliva	20	Undigested and waste products expelled from the body
U	Stomach	21	Controls the passing of faeces from the body
V	Villi	22	An enzyme that digests starch to maltose

Chapter 14:
Useful microbes

Microbes (fungi, bacteria, viruses and moulds) can be useful

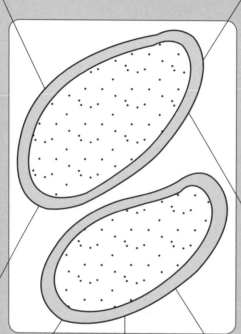

SEWAGE TREATMENT

Decay bacteria carry out AEROBIC DIGESTION of solutes dissolved in water

COMPLEX COMPOUNDS → CO_2, H_2O, H_2S and ammonium compounds

nitrifying ↓ bacteria

NITRATES

ANAEROBIC DIGESTION

COMPLEX COMPOUNDS → Fatty acids, amino acids and sugars

Also produce METHANE (in BIOGAS), an important fuel

FOODS

- YOGHURT: *Lactobacillus* excretes lactic acid by fermenting lactose (milk sugar) – the acid thickens milk and gives it a sour taste
- VINEGAR: a solution of acetic acid made when *Acetobacter* ferments alcohol
- MYCOPROTEIN: the bodies of the fungus *Fusarium* are dried and squeezed

SOURCE OF ENZYMES

- LIPASE – used to digest fatty/greasy stains when part of a BIOLOGICAL WASHING POWDER
- PROTEASE – used to soften leather and remove stains from false teeth
- AMYLASE – used to digest starch for cakes and biscuits

VACCINES

- Some are produced in genetically modified yeast
- 'WEAKENED' bacteria can be used as vaccines themselves

ANTIBIOTICS

Moulds (e.g. *Penicillium*) can be grown in bioreactors to produce compounds (e.g. penicillin) that can control the reproduction of harmful bacteria

GENETIC ENGINEERING

Bacteria can be genetically modified to produce proteins (e.g. insulin) which are useful to humans

VIRUSES can carry helpful genes to repair damage: GENE THERAPY

BREWING AND BAKING

Yeast

GLUCOSE ⟶ ALCOHOL + CARBON DIOXIDE

Anaerobic conditions

is poisonous, and eventually kills the yeast!

'fizz' in some drinks (e.g. beer)

'bubbles' that make bread rise

Selective breeding and genetic engineering

Selective breeding and genetic engineering are methods we can use to produce plants and animals better suited to our purposes.

SELECTIVE BREEDING

For hundreds of years farmers have been breeding animals and plants to give greater productivity. By selecting parents with desired characteristics and mating them, there is a better chance of the offspring having those characteristics. For example, if a champion stallion (male horse) is mated with a champion mare (female horse) the foals (young horses) are likely to be fast runners.

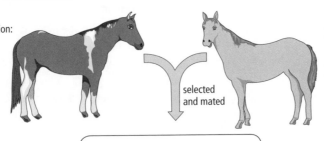

champion stallion:
fast
large muscles
healthy
good stamina

selected
and mated

champion mare:
fast
large muscles
healthy
good stamina

offspring likely, but not certain, to inherit parents' characteristics

Selective breeding has been very successful but it is slow and unpredictable.

GENETIC ENGINEERING

Genetic engineers can identify single genes in the DNA of a plant or animal. They can then 'cut' that gene out and introduce it into other cells for duplication. The example below shows how genetic engineering can make insulin for use by people with diabetes.

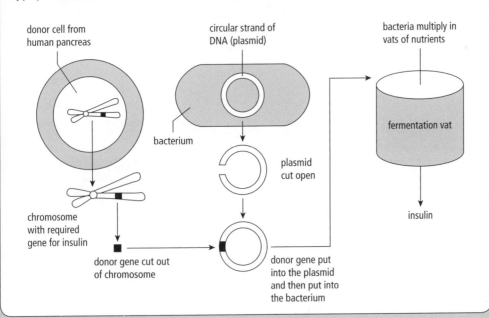

donor cell from human pancreas

circular strand of DNA (plasmid)

bacteria multiply in vats of nutrients

bacterium

fermentation vat

plasmid cut open

chromosome with required gene for insulin

insulin

donor gene cut out of chromosome

donor gene put into the plasmid and then put into the bacterium

POTENTIAL USES OF GENETIC ENGINEERING

- Production of important medical products, e.g. insulin, human growth hormone, blood clotting agent (factor 8)
- Gene therapy for inherited diseases, e.g. cystic fibrosis
- Plants which can resist frost and disease, or which can produce bigger fruit with more flavour
- Animals which produce more milk or meat and which are resistant to common diseases

1. Cheese is made using milk. The milk can be supplied by many different animals. The table shows the composition of three different types of milk.

SOURCE OF MILK	MASS OF SUBSTANCE/G PER 100 G MILK				
	Carbohydrate	Fat	Protein	Minerals	Water
Sheep	5.0	9.5	4.5	1.0	80.0
Goat	5.1	6.0	3.5	0.7	84.7
Cow	5.1	3.5	3.0	0.8	87.6

a. To make soft cheese the whey (a liquid containing 50 per cent of the water and all of the carbohydrate) is removed. 100 g of cow's milk makes 51.1 g of soft cheese. How much soft cheese would be made from 100 g of goat's milk? Show your working. (2)

b. Each gram of carbohydrate can be respired to 18 kJ of energy. What would be the energy content of the whey separated from 100 g of sheep's milk? Show your working. (2)

c. The flow chart shows a scheme for the industrial production of cheese. Match the labels to the stages of the process.

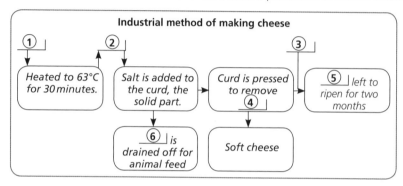

Industrial method of making cheese

A: bacteria break down proteins and fats
B: addition of chymosin or rennin
C: water
D: whey
E: hard cheese
F: fresh milk (6)

2. The industrial production of many proteins is made easier by the use of bioreactors such as the one shown in the diagram below.

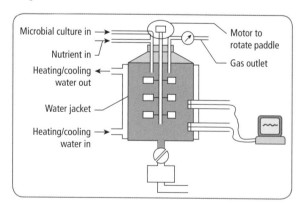

a. Why is it important that the water jacket keeps the temperature close to 30°C? (2)

b. What is the function of the paddles? (1)

c. An experiment was carried out to find the best conditions for producing the protein. Samples of the liquid inside the fermenter were taken at five hour intervals, and analysed for the protein product. The results of the experiment are shown in the table below.

TIME/HOURS	MASS OF PRODUCT/KG
5	2.12
10	2.50
15	2.80
20	3.04
25	3.28
30	3.40
35	3.50
40	3.50
45	3.50

i. Plot these results in the form of a graph (4)

ii. The company collects the protein as soon as the yield reaches 90 per cent of its maximum. At what time can the product be collected? Show your working. (3)

3. The diagram shows some of the stages involved in making beer.

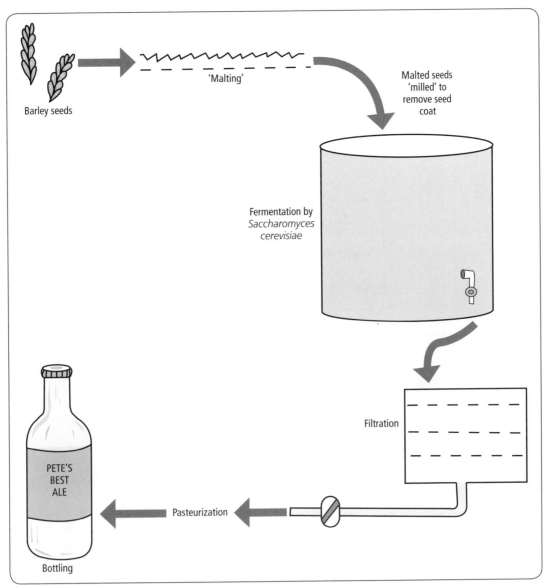

a. During the malting stage the barley seeds germinate. Starch in the seeds is converted to maltose.

   i. Suggest two environmental conditions required for malting. (2)

   ii. Name the enzyme responsible for converting starch to maltose. (1)

b. Name the genus of the organism responsible for the fermentation process. (1)

c. The rate of alcohol production depends on the temperature in the fermentation vat. The results in the table were collected from an experiment designed to test this effect of temperature.

TEMPERATURE/°C	10	20	30	40	50	60
RATE OF ALCOHOL PRODUCTION/ UNITS OF ALCOHOL PER HOUR	0.5	1.4	2.7	3.5	2.3	0.0

   i. Plot a graph of this information. (4)

   ii. Calculate the increase in alcohol production between 25 and 35°C. (1)

   iii. Explain the result of heating the mixture to 60°C. (2)

d. Suggest why the beer is pasteurized before it is bottled. (1)

e. Name the gas that gives the 'fizz' to bottled beer. (1)

4. The diagram shows a fermenter that may be used to commercially produce an antibiotic. The culture medium is infected with an aerobic antibiotic-producing mould.

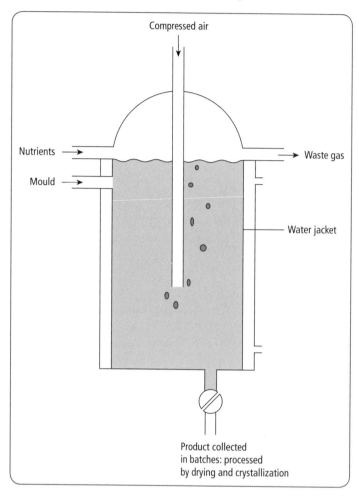

Compressed air

Nutrients →

Mould →

→ Waste gas

Water jacket

Product collected in batches: processed by drying and crystallization

a. i.   What is meant by the term 'aerobic'? (1)
   ii. Why is it necessary to supply the mould with glucose? (1)
   iii. Suggest one reason why the antibiotic is dried and crystallized once the contents of the fermenter have been collected. (1)
b. Name one antibiotic that could be produced in this system. (1)
c. Sore throats are sometimes caused by a bacterium called *Streptococcus*. This type of sore throat can be treated with an antibiotic, but there is a danger that the bacterium can develop a resistance to the antibiotic.
   The diagram shows how an antibiotic-resistant form may develop.

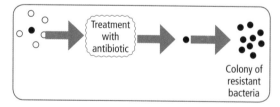

Treatment with antibiotic

Colony of resistant bacteria

   i.   Name the process that leads to the **formation** of the antibiotic-resistant bacterium. (1)
   ii. Name the process that leads to the development of a colony of antibiotic-resistant bacteria while killing off the non-resistant bacterial cells. (1)
d. Explain the difference between an antiseptic and an antibiotic. (2)

5. The diagram shows how human factor 8 can be made by genetic engineering.

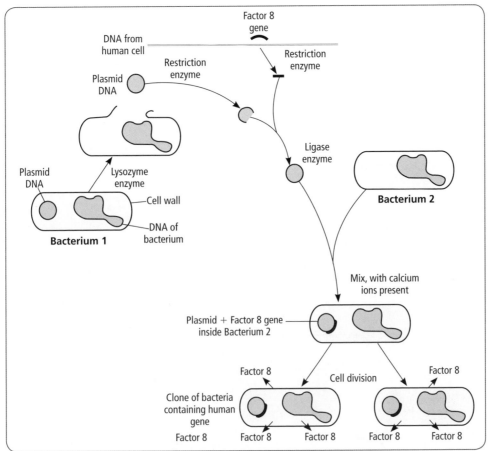

a. Using only the information from the diagram describe how the gene for factor 8 is put into a bacterium. (7)

b. Why is factor 8 valuable? (1)

c. Name one other human protein that can be made by genetic engineering, and state why it is so useful. (2)

d. A bacterium that has been altered in this way is a genetically modified organism (GMO). Why are some people against the use of GMOs? (2)

6. Antibiotics are useful drugs. The antibiotic, amoxycillin, can be manufactured by growing a mould in a nutrient solution in a fermenter.

The graph shows how the concentration of the nutrient changes over time, in a fermenter.

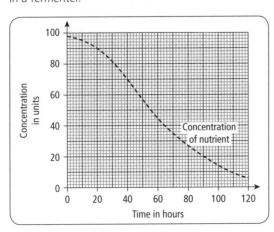

a. The table shows how the concentration of amoxycillin changes in the fermenter.

TIME, IN HOURS	0	20	40	60	80	100	120
CONCENTRATION OF AMOXYCILLIN, IN UNITS	0	1	57	86	93	98	99

On the grid above, draw the graph for amoxycillin production. (2)

b. Explain why the nutrient concentration in the fermenter changes over time. (1)

c. Describe the relationship between the concentration of nutrient and the concentration of amoxycillin. (2)

d. Why do doctors give their patients antibiotics? (1)

7. The diagram shows how genetic engineering can be used to produce human insulin from bacteria. Ampicillin and tetracycline are two types of antibiotic. Study the diagram carefully and answer the questions.

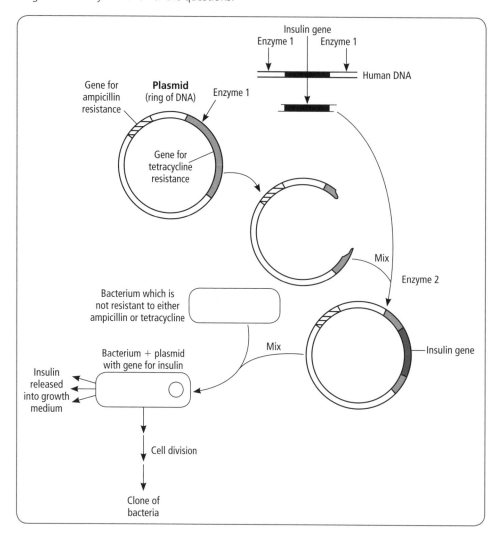

In experiments like these, some bacteria take up the plasmid (ring of DNA) containing the insulin gene. Other bacteria fail to take up a plasmid, or they take up an unmodified plasmid (a ring of DNA which has not been cut open and which does not contain the insulin gene).

a. Complete the table by putting a tick (✓) in the correct boxes to show which bacteria would be able to multiply in the presence of ampicillin and which bacteria would be able to multiply in the presence of tetracycline.

b. The bacterium with the plasmid containing the insulin gene multiplies by cell division to form a clone of bacteria.

    Will **all** the bacteria in this clone be able to produce insulin? Explain your answer.　(3)

	BACTERIUM CAN MULTIPLY IN THE PRESENCE OF	
	AMPICILLIN	TETRACYCLINE
Bacterium + plasmid with the insulin gene		
Bacterium without a plasmid		
Bacterium with an unmodified plasmid		

(3)

USEFUL MICROBES: Crossword

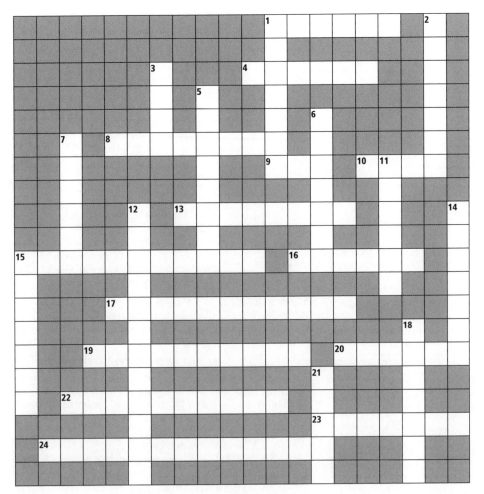

ACROSS:
1 A useful fuel released when waste materials decompose
4 A mixture of faeces, water and bacteria: must be treated before the water is safe to drink
8 Enzyme, produced by genetic engineering, that is used in cheese production
9 Abbreviation for a genetically modified organism
10 With 7 down: responsible for the acidic taste of yoghurt
13 Most widely used organisms in biotechnology, can act as 'factories' to produce proteins using human genes as instructions
15 Dried hyphae of a fungus – a healthy alternative to meat?
16 A means of carrying a gene from one organism to another
17 Enzyme used in genetic engineering to cut into DNA molecules
19 Important drug, produced by molds, that may kill bacteria or slow down their rate of multiplication
20 Component of biological washing powders-an enzyme that helps to remove greasy stains
22 The tasty component of vinegar, produced by fermentation with *Acetobacter*
23 Toxic product of anaerobic respiration in 21 down
24 Bacterium that ferments the sugar in milk

DOWN:
1 Industry that produces 23 across
2 A small piece of bacterial DNA that can carry a gene into a bacterial cell
3 The liquid part of soured milk, removed during production of cheese
5 Important enzyme in the leather-softening process
6 With 11 down – the gaseous product of anaerobic respiration (puts the 'fizz' into many drinks)
7 See 10 across
11 See 6 down
12 Set of reactions carried out by microbes with many products useful to humans
14 Enzyme used to break down starch in the early stages of 1 across
15 Can be a useful product for powering machinery at landfill sites (although it is also a greenhouse gas)
18 A useful addition to petrol in internal combustion engines
21 Type of microbe responsible for alcoholic fermentation

REVISION SUMMARY: MATCHING TERMS

Match the words from the list below with the definitions in the table.

WHEY, FERMENTATION, YEAST, LACTOBACILLUS, CARBON DIOXIDE, ALCOHOL, MYCOPROTEIN, BACTERIUM, CHYMOSIN, ACETOBACTER, BIOREACTOR, PASTEURIZATION, STERILIZATION, DISTILLATION, HOPS

DEFINITION	MATCHING WORD
Increases proportion of alcohol by removal of water	
Anaerobic respiration	
Vessel in which growth of microbes can be controlled for industrial or medical reasons	
Bacterium that plays an important part in the production of cheese	
Gassy product of yeast fermentation – important in fizzy wines and beers	
Liquid by-product of cheese manufacture	
Add the flavour to beer	
The key organism in both brewing and baking	
Fermentation product that is useful in medicine but may lead to intoxication	
Milk treatment that kills all harmful microbes	
Important enzyme in the formation of solid cheeses	
Bacterium involved in the production of vinegar	
Treatment that kills all microbes in a food or medical product	
Healthy product from the bodies of the fungus *Fusarium graminearum*	
Important microbe to humans, even though it has no nucleus	

Plants make food by photosynthesis

LEAVES ARE ADAPTED FOR PHOTOSYNTHESIS

Leaves are **thin**: few layers of cells so light can reach the sites of photosynthesis.

Large **surface area** to trap the maximum amount of sunlight.

Vascular bundles transport water and mineral salts in (**xylem**) and sugars out (**phloem**).

Palisade layer near in upper surface has many chloroplasts in cells tightly packed for maximum light absorption.

Spongy layer has loosely packed cells, covered with a thin film of water to allow rapid diffusion of carbon dioxide.

Stomata are pores that allow the entry of **carbon dioxide** and the exit of **oxygen**. The pores can be opened and closed by the action of guard cells. This conserves water because water easily evaporates from the surface of spongy cells.

The occurrence of photosynthesis can be demonstraled by testing for the presence of starch (made from glucose).

$$STARCH \xrightarrow{IODINE\ SOLUTION} INTENSE\ BLUE\text{-}BLACK\ COLOURATION$$

The rate of photosynthesis can be measured by collecting evolved oxygen, usually from the cut end of the aquatic plant, *Elodea*.

EQUATION FOR PHOTOSYNTHESIS

$$CARBON\ DIOXIDE + WATER \xrightarrow[\substack{CHLOROPHYLL\\ ENZYMES}]{LIGHT\ ENERGY} GLUCOSE + OXYGEN$$

$$6CO_2 + 6H_2O \longrightarrow C_6H_{12}O_6 + 6O_2$$

PLANTS NEED MINERALS TOO!

MINERAL	SOURCE	IMPORTANCE
Magnesium	Absorbed as Mg^{2+} from the soil solution.	Manufacture of chlorophyll: absence makes leaves turn yellow and eventually stops photosynthesis.
Nitrogen	Absorbed as nitrate (NO_3^-) or ammonium (NH_4^+) from the soil solution.	Manufacture of proteins, nucleic acids and plant hormones: absence causes poor growth, especially of leaves. Is an important limiting factor, so many farmers add nitrogen-containing fertilizers to their land.

APHIDS TAP PHLOEM FOR THE SUGARS MADE BY PHOTOSYNTHESIS

YUMMY!

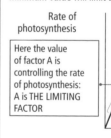

LIMITING FACTORS AND PHOTOSYNTHESIS

There are a number of conditions that must be satisfied for photosynthesis to proceed. The factor that **is nearest to its minimum value** will limit the rate of this process.

Rate of photosynthesis

Here the value of factor A is controlling the rate of photosynthesis: A is THE LIMITING FACTOR

Increase in availability of A does not increase rate of photosynthesis there is some other LIMITING FACTOR

Availability of factor A

Limiting factors in photosynthesis are **carbon dioxide concentration light intensity temperature** – affects the enzymes that catalyse the chemical reactions of photosynthesis.

WHAT HAPPENS TO THE GLUCOSE?

Glucose

- Respiration to provide the energy to drive the metabolic reactions needed to keep the plant alive.
- Conversion to other molecules such as oils and proteins. This may require mineral salts (e.g. nitrate).
- Conversion to sucrose for transport to other parts of the plant via the phloem.
- Conversion to cellulose for the construction of plant cell walls.
- Conversion to starch for storage. Starch is insoluble so does not affect water potential of plant cells.

1. An experiment was carried out using three plants of the same species. Each one was kept in a dark cupboard for 12 hours before the experiment was carried out.

One leaf was detached from each plant and immediately placed in test-tubes, labelled **A**, **B** and **C** as shown in the diagram below. The tubes were then closed using bungs.

Red sodium hydrogencarbonate indicator solution was placed in each of the three tubes.

The indicator is

red in normal air;

purple when there is less carbon dioxide than in normal air;

yellow when there is more carbon dioxide than in normal air.

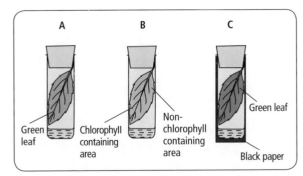

The test-tubes were then left in bright sunlight for 3 hours. The colour of the indicator solution was found to have changed in some of the test-tubes.

a. i. Suggest the colour of the indicator in each test-tube after 3 hours.

ii. Explain each answer given in **(a) (i)**.

b. i. Describe how a green leaf may be tested to show the presence of starch. State what safety precautions should be taken. (4)

ii. Copy the diagram below and shade the area of each leaf which would show the presence of starch.

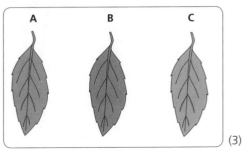

(3)

CIE 0610 June '98 Paper 6 Q2

2. The diagram shows the apparatus used to investigate the effect of light intensity on the rate of photosynthesis.

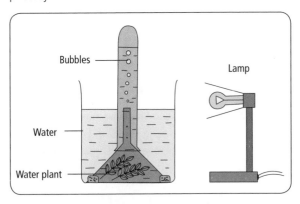

a. How would the light intensity be varied? (1)
b. Suggest a suitable control for the investigation. (1)
c. Name two factors that should be kept constant if this is to be a fair test. (2)
d. The rate of photosynthesis was measured by counting the number of bubbles released into the test tube every 5 minutes. The results obtained are shown below.

LIGHT INTENSITY / ARBITRARY UNITS	RATE OF PHOTOSYNTHESIS / NUMBER OF BUBBLES PER 5 MINUTE PERIOD
0	1
10	8
20	15
30	23
40	30
50	37
60	43
70	44
80	44

i. Plot a graph of these results. (5)
ii. Use your graph to find the number of gas bubbles given off in a five minute period at a light intensity of 35 units. (2)
iii. Describe the relationship between light intensity and rate of photosynthesis. Explain this relationship. (4)
iv. Suggest **two** ways in which the bubbles of gas differ from normal atmospheric air. (2)
v. Suggest one way in which the results of the experiment could be made more reliable. (1)

e. Two students remembered that the school aquarium had red-coloured lighting. They were interested in whether the colour (wavelength) of light would affect the rate of photosynthesis. Describe an experiment that they could carry out to test the hypothesis that 'colour of light affects the rate of photosynthesis'. Name the input variable, the outcome variable and any fixed variables in their experiment. (5)

3. The graph shown below indicates the amount of sugar contained in the leaves of a group of plants, kept in a greenhouse, over a seven-day period.

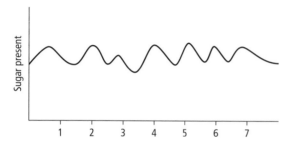

a. Write a word equation for the process that produces the sugar in the plants. (4)
b. Name the process shown by the equation. (1)
c. Why does the sugar level rise and fall during the day? (2)
d. Suggest a reason for the lower peak on day 3. (2)
e. The greenhouse owner used a paraffin burner during the spring. Give two reasons why using the burner might increase the yield of sugar produced. (2)

4. The diagram shows a section through part of a leaf.

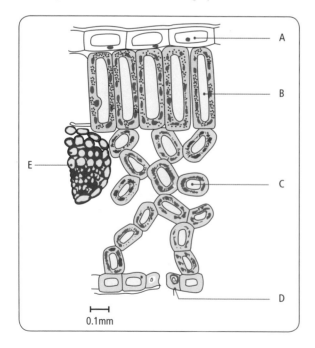

0.1mm

a. i. Name the cells labelled A, B, C, D and E. (5)
   ii. In which of these cells does most photosynthesis occur? (1)
   iii. Which of these cells would transport water and minerals to the leaf? (1)
b. In what form is carbohydrate transported around the plant? (1)
c. i. In what form is carbohydrate stored in an onion? (1)
   ii. Describe how you would carry out a test for this storage carbohydrate. Describe a positive result. (2)
   iii. Describe two uses, other than storage, for the carbohydrate made by the process of photosynthesis. (2)
d. Look back at the leaf section. Use the scale to work out
   i. the thickness of the leaf
   ii. the length of a palisade cell
   Show your working. (2)

5. Three plants were grown to study the effects of nitrate and magnesium ion deficiency on their development. They were kept in the same conditions, except for the types of minerals supplied.
Plant **A** was provided with all essential minerals.
Plant **B** was given all minerals except nitrate ions.
Plant **C** was given all minerals except magnesium ions.
The diagram shows the plants a few weeks later.

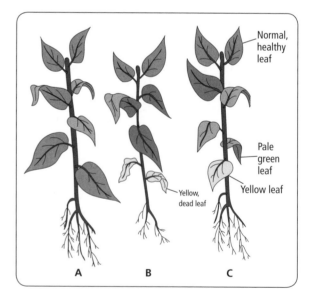

a. State three conditions, **other than** water and the concentration of mineral ions, that would need to be kept the same for all the plants, in order to make the investigation a fair test. (3)
b. Describe and explain the effect on plant growth of
   i. a deficiency of nitrate ions on plant **B**;
      description
      explanation (4)

ii. a deficiency of magnesium ions on plant **C**.
description
explanation (2)

c. A farmer tested the soil in a field and found that
there was a high nitrate ion concentration.
The farmer then grew a crop in this field.
After the crop was removed, the soil was tested
again. The nitrate ion concentration had decreased.
i. Suggest two reasons why the nitrate ion
concentration had decreased. (2)

ii. Describe two methods the farmer could use to
improve the nitrate ion concentration in
the soil. (2)

d. Some species of plant grow well in soil that is
always low in nitrate ions.
Explain how they can obtain a source of nitrogen
compounds. (3)

CIE 0610 June '05 Paper 3 Q1

REVISION SUMMARY: Match these terms to their definitons

	TERM		DEFINITION
A	Limiting factor	1	The molecule that absorbs light energy to drive photosynthesis
B	Nitrate	2	The form in which carbohydrate is transported away from the leaves
C	Magnesium	3	Are needed to catalyse the chemical reactions in photosynthesis
D	Chlorophyll	4	Some aspect of the environment needed for photosynthesis that is nearest to its minimum value
E	Palisade cell	5	Gas that is a raw material for photosynthesis
F	Stomata	6	The plant tissue that transports the products of photosynthesis away from the leaves
G	Carbon dioxide	7	The organelle in which photosynthesis takes place
H	Oxygen	8	Pores that allow the entry of carbon dioxide
I	Light	9	A molecule made from glucose that is essential for plant cell structure
J	Phloem	10	Mineral ion required for protein synthesis
K	Sucrose	11	Storage carbohydrate in plants
L	Cellulose	12	The source of energy for photosynthesis
M	Starch	13	Adapted for its function by being packed with chloroplasts
N	Chloroplast	14	Gaseous product of photosynthesis essential for aerobic respiration
O	Enzymes	15	Mineral ion essential for the production of chlorophyll

WATER UPTAKE BY PLANTS

can be measured using a bubble potometer.

Bubble

Water loss draws bubble along capillary tube as water uptake replaces loss.

Water Movement through a plant

begins with evaporation from the leaf surface.

Transpiration is evaporation of water from mesophyll cells followed by loss of water through the stomata. This occurs by evaporation and lowers the water potential in the tissues of the leaf.

Stomata are essential if the uptake of CO_2 for photosynthesis is to go on, but if stomata are open water evaporated from the spongy mesophyll can diffuse out of the leaf down the gradient of water potential.

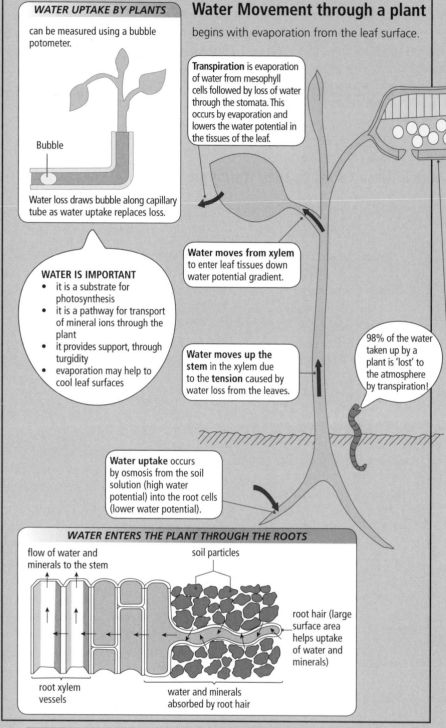

WATER IS IMPORTANT

- it is a substrate for photosynthesis
- it is a pathway for transport of mineral ions through the plant
- it provides support, through turgidity
- evaporation may help to cool leaf surfaces

Water moves from xylem to enter leaf tissues down water potential gradient.

Water moves up the stem in the xylem due to the **tension** caused by water loss from the leaves.

98% of the water taken up by a plant is 'lost' to the atmosphere by transpiration!

LEAF STRUCTURE MAY REDUCE TRANSPIRATION

- thick, waxy cuticle reduces evaporation from epidermis.
- stomata may be sunk into pits, which trap a pocket of humid air.
- leaves may be rolled with the stomata on the inner surface close to a trapped layer of humid air.

HUMID AIR

- leaves may be needle-shaped to reduce their surface area.

Water uptake occurs by osmosis from the soil solution (high water potential) into the root cells (lower water potential).

ATMOSPHERIC CONDITIONS MAY AFFECT TRANSPIRATION

- **Wind** moves humid air away from the leaf surface and increases transpiration.
- **High temperatures** increase the water holding capacity of the air and increase transpiration.
- **Low humidity** increases the water potential gradient between leaf and atmosphere and increases transpiration.
- **High light intensity** causes stomata to open (to allow photosynthesis), which allows transpiration to occur.

WATER ENTERS THE PLANT THROUGH THE ROOTS

flow of water and minerals to the stem

soil particles

root hair (large surface area helps uptake of water and minerals)

root xylem vessels

water and minerals absorbed by root hair

TRANSPIRATION IS CONTROLLED BY STOMATA

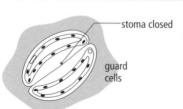

stoma open

guard cells

stoma closed

guard cells

Guard cells swollen with water.
Stoma (pore) is held open.
Water can be lost through the stoma.

Guard cells limp with lack of water.
Stoma (pore) is closed.
Little water is lost through the stoma.

Transport systems in plants

move vital substances from **sources** (sites of uptake or manufacture) to **sinks** (sites of production or storage).

INVESTIGATION OF FUNCTIONS OF PLANT TISSUES
- **Phloem** aphids (greenfly) can sample contents and radioactive sugars identify tissue as site of sugar transport.
- **Xylem** water-soluble dyes trace pathway of water movement
- **Roots** inhibitors of respiration stop active uptake of ions

FRUITS AND GROWING POINTS these are **sinks** for many nutrients.

Fruits
- demand water for swelling of ovary wall if succulent
- demand sucrose to be converted to starch as an energy store

Growing points
- demand water for cell swelling
- demand sucrose as energy source for cell division
- demand all nutrients as raw materials for cell production.

STEM the position of the vascular bundles (in a ring with soft cortex in the centre) helps to support the stem against sideways forces, e.g. wind.

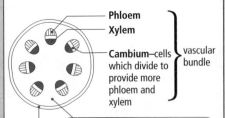

Phloem
Xylem
Cambium—cells which divide to provide more phloem and xylem

} vascular bundle

Cortex cells become turgid (filled with water) and help to support non-woody parts.

Epidermis protective against, for example, infection by viruses and bacteria, and dehydration.

LEAVES are both **sinks** and **sources**.

Sinks	Sources
• water as a reactant in photosynthesis	• glucose formed during photosynthesis
• magnesium as a component of the chlorophyll molecule	

CARBON DIOXIDE + WATER
↓
GLUCOSE + OXYGEN

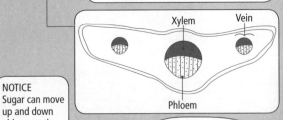

Xylem   Vein

Phloem

ROOT

Root hair extended cells of epidermis increase surface area for water/ion uptake.

Epidermis protective, e.g. against infection by fungi.

Phloem
Xylem

} together form a strong central rod which helps to prevent root being pulled out of the soil.

Cortex (pith) can act as a winter store for starch.

PHLOEM transport of organic products of photosynthesis i.e. sugars (carried as sucrose) and amino acids.

XYLEM transport of water and dissolved mineral salts - movement is always **up** the stem.

NOTICE Sugar can move up and down phloem at the same time

DIRECTION OF TRANSPORT VARIES WITH THE SEASONS! Sucrose will be transported **from** stores in the root **to** leaves in spring, but **to** stores in the root **from** photosynthesising leaves in the summer and early autumn.

ROOTS: are both **sinks** and **sources**.

Sinks	Sources
• sucrose to supply energy for growth and active uptake of ions from the soil	• water, absorbed from the soil solution
• sucrose to be converted to starch for storage	• ions, absorbed from soil by active transport
	• sucrose, before leaves are capable of photosynthesis, in spring.

1. Using the apparatus shown below, a student set out to investigate the factors that influence water uptake by a plant.

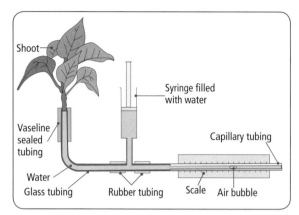

After a series of experiments he obtained the following results.

ENVIRONMENTAL CONDITION	TIME TAKEN FOR BUBBLE TO MOVE 10 CM / MIN	RATE OF BUBBLE MOVEMENT / CM PER MINUTE
1 High light intensity	5	
2 High humidity (plant enclosed in clear plastic bag)	16	
3 Wind (electric fan blowing over plant surface)	1	
4 Dark and windy	17	
5 Dark and low humidity	25	

a. Calculate the **rate** at which the bubble moves. Plot these results in the form of a bar chart. (5,5)

b. Plants require light for photosynthesis, and 'anticipate' the need for carbon dioxide uptake during photosynthesis by opening their stomata under appropriate conditions. Does this help to explain the results of the first experiment? Explain your answer. (2)

c. Water is lost from leaves by evaporation and diffusion **if the stomata are open, and a suitable water potential gradient exists**. Use this information to explain the results of experiments 2 and 5. (3)

d. How can you explain the results of experiments 3 and 4? (2)

2. Four leaves were removed from the same plant. Petroleum jelly (a waterproofing agent) was spread onto some of the leaves, as follows:

Leaf **A**: on both surfaces
Leaf **B**: on the lower surface only
Leaf **C**: on the upper surface only
Leaf **D**: none applied

Each leaf was then placed in a separate beaker, as shown in the diagram.

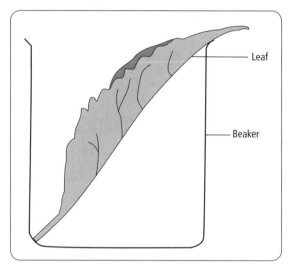

Each beaker was weighed at intervals. The results are shown in the graph.

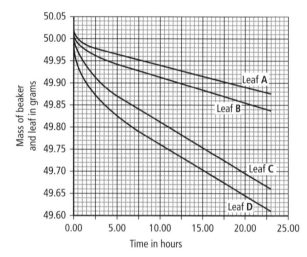

a. Give evidence from the graph in answering the following questions.
i. Which surface (upper or lower) loses water most rapidly? (1)
ii. Is water lost from both surfaces of the leaf? (1)

b. The diagrams below show the appearance of each surface of the leaf as seen through a microscope.

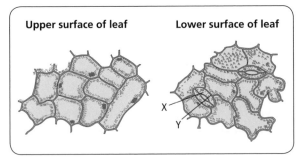

Upper surface of leaf	Lower surface of leaf

i.   Name space **X** and cell **Y**. (2)
ii. Use information in diagram **2** to explain why the results are different for leaves **B** and **C**. (2)

3. The diagrams show the underside of two leaves.

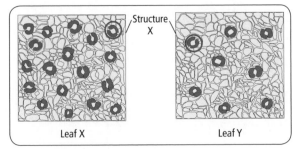

Leaf X   Leaf Y

a. Name the structures labelled A. (1)
b. Make a labelled drawing of a side view of one of these structures. (3)
c. Water can be lost from a leaf through these structures. The water on the surface of spongy mesophylll cells changes to water vapour, and then moves out of the leaf down a concentration gradient.
   i.   Name the process in which water changes from liquid to vapour. (1)
   ii. Name the process by which the molecules of water vapour move down a concentration gradient. (1)
   iii. Suggest one advantage and one disadvantage of this process to the plant. (2)
d. The view of the leaf underside covers 1 mm². Each of the leaves measured 5 cm × 2 cm. Calculate the total number of the structures A on each of the leaves. Show your working. (4)

LEAF	NUMBER OF STRUCTURES A IN 1 MM²	TOTAL NUMBERS OF STRUCTURES A ON LEAF
X		
Y		

e. Which of the leaves is better adapted to living in dry, hot conditions? Explain your answer. (2)
f. Give two other features of leaves adapted to life in dry, hot conditions. (2)

4. The diagram on the left below shows a transverse section through an *Ammophila* leaf. This plant has very long roots. The diagram on the right shows a cactus plant. Both plants live in very dry conditions.

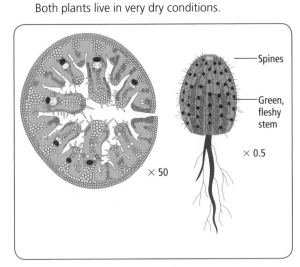

a. Suggest how each of the following adaptations would enable the named plant to survive in very dry conditions.
   i.   *Ammophila*
      1. rolled leaves with stomata on the inside of the leaf (2)
      2. thick waxy cuticle on the outside of the leaf (1)
   ii. Cactus
      1. very long roots (1)
      2. fleshy green stem (2)
b. Suggest why having only a few, very small leaves could be a disadvantage to a plant. (2)
c. Water is involved in a number of processes in plants.
   Complete the table by
   i.   naming the processes described;
   ii. stating one variable that, if increased, would speed up the process.

DESCRIPTION OF PROCESS	NAME OF PROCESS	VARIABLE THAT, IF INCREASED, WOULD SPEED UP THE PROCESS
Absorption of water from the soil		
Using water to form glucose		
Movement of water vapour out of leaves		

(6)

CIE 0610 June '05 Paper 3 Q4

REVISION SUMMARY: Fill in the missing words

Use words from the following list to complete the paragraphs below.

PHLOEM, VASCULAR, ACTIVE TRANSPORT, OSMOSIS, DIFFUSION, RESPIRATION, XYLEM, SURFACE AREA, NITRATE, IONS, SOLVENT, PHOTOSYNTHESIS, DIGESTION, SUPPORT, HAIRS, EPIDERMIS

Water is obtained by plants from the soil solution. The water enters by the process of ...., via special structures on the outside of the root called root .... These structures increase the .... of the root and, as well as absorbing water, they can also take up .... such as ...., which is required for the synthesis of chlorophyll. These substances are absorbed both by .... and by ...., a process that requires the supply of energy.

Plant cells rely on water for ...., as a .... and as a raw material for .... Water is used as a transport medium for both ions and sugars. The ions are transported, along with water, in the .... tissue. Sugars are transported through living cells of the .... and these specialized tissues are grouped together into .... bundles. (13)

Blood and the human circulatory system

Blood tissue transports vital substances around the body.

It also plays an essential part in protecting the body from damage and disease.

BLOOD TISSUE

Spinning blood in a centrifuge separates the plasma from the cells.

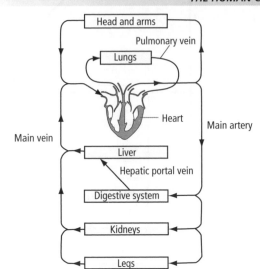

Plasma (55%)

Cells (45%)

Water
+
Dissolved chemicals
+
Plasma proteins

Red cells
+
Lymphocytes
+
Phagocytes
+
Platelets

THE FUNCTIONS OF BLOOD CELLS

Red blood cells carry oxygen by combining it with **haemoglobin**.

Lymphocytes fight disease by producing **antibodies** that destroy dangerous microbes or mark them for attack by phagocytes.

Phagocytes fight disease by **engulfing** dangerous microbes.

Platelets are cell fragments. They help **clotting**, preventing blood loss and infection.

THE HEART

Blood is pumped around the body by a muscular pump called the **heart**.

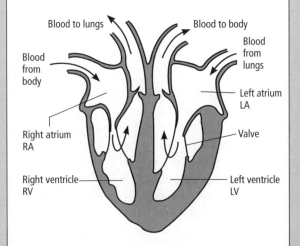

Blood to lungs

Blood to body

Blood from body

Blood from lungs

Right atrium RA

Left atrium LA

Valve

Right ventricle RV

Left ventricle LV

- Blood in the RV is pumped to the lungs where it is oxygenated.
- Blood from the lungs flows back into the LA and then into the LV.
- Blood in the LV is pumped through the body (except for the lungs).
- Blood returns to the heart where it enters the RA.

Blood passes through the heart twice for each trip around the body. We call this **the double circulation**.

Valves in the heart prevent blood from being pushed backwards up into the atria when the heart 'beats'.

THE HUMAN CIRCULATORY SYSTEM

Head and arms

Pulmonary vein

Lungs

Heart

Main vein

Main artery

Liver

Hepatic portal vein

Digestive system

Kidneys

Legs

- **Arteries** carry blood **away** from the heart. Blood in arteries has more oxygen than blood in veins – except for the pulmonary artery and veins, which go to and from the lungs!
- **Arteries** have thick walls to withstand higher blood pressure.
- **Veins** have thinner walls but have valves to stop blood flowing 'backwards'.
- **Veins** carry blood back to the heart.
- **Capillaries** reach all tissues and organs.

1. The table shows the cell composition of three samples of blood.

	JACK	JAMES	JULIAN
Red cells / number per mm³	8 200 000	5 000 000	2 200 000
White cells / number per mm³	500	8000	5000
Platelets / number per mm³	280 000	255 000	1000

a. Which person is most likely to have recently lived at high altitude? Explain your answer. (2)

b. Which person would be least likely to resist infection by a virus? Explain your answer. (2)

c. People at risk from heart disease are sometimes recommended to take half an aspirin each day. Aspirin reduces the clotting of blood, and helps to prevent blood clots blocking narrowed arteries. Which person is most likely to be taking this drug each day? Explain your answer. (2)

d. Iron deficiency in the diet can cause a condition called anaemia. Which person is likely to show the symptoms of anaemia? Explain your answer. (2)

e. These three samples were all taken from 25-year-old men. Explain why this makes comparisons between them more valid. (2)

2. The diagram below shows three types of blood vessel in the human circulation.

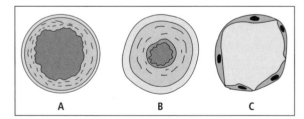

A   B   C

Use the label letters (A, B or C) to identify the vessels that

a. have walls one cell thick

b. allow amino acids to cross the walls

c. have valves to prevent back-flow

d. carry blood to the heart

e. carry blood under high pressure

f. pulse as blood flows through them

g. can be blocked by fatty tissue called atheroma

h. carry blood away from the heart

i. have thick elastic walls

j. increase greatly in number after long periods of athletic training (10 × 1)

3. Examine this bar chart.

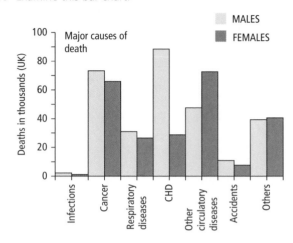

a. How many deaths in the UK result from CHD? (1)

b. How many deaths result from CHD and other circulatory diseases combined? (2)

c. What proportion of all deaths in the UK is due to CHD? (2)

d. Redraw the graph in the form of a pie chart. Which of the two – bar chart or pie chart – gives the information from **c** most clearly? Explain your answer. (5)

4. The diagram below shows part of the human circulation.

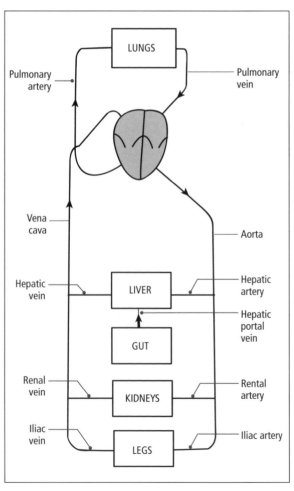

a. Name the blood vessels and chambers of the heart through which
   i. a molecule of amino acid absorbed by the gut passes on its way to build up muscle in the heart (3)
   ii. a red blood cell passes as it delivers oxygen from the lungs to the liver and returns to be reoxygenated. (3)
b. A poster in a doctor's surgery had the headline

   A HEALTHY LIFESTYLE MEANS A HEALTHY HEART

   i. Suggest **two** features of a person's lifestyle that would give them a good chance of having a healthy heart. (2)
   ii. Suggest two **different** features of a person's lifestyle that might lead to heart disease. (2)

5. The graph below shows pressure changes, measured in millimetres of mercury, that occur during one complete beat of a human heart. **Data for pressures In the right atrium and the right ventricle are not shown**.

   Look at the graph and answer the following questions.

a. i. How long does one complete beat of the heart last? (1)
   ii. Calculate the number of beats per minute. **Show your working** (2)
b. i. The maximum pressure reached by the left ventricle is five times that reached by the right ventricle.
      Using the data in the graph calculate the maximum pressure reached in the right ventricle.
      Show your working (1)
   ii. From your **knowledge of the structure of the heart**, explain how this difference in pressure between the ventricles is achieved. (1)
c. Explain the connection between the pressure in the left ventricle and the pressure in the aorta between 0.3 and 0.5 seconds. (1)
d. From your **knowledge of how the heart works,** suggest which valves close at points X and Y. In each case explain why the valve closes at that point.
   Name of valve closing at X
   Reason for closing
   Name of valve closing at Y
   Reason for closing (4)

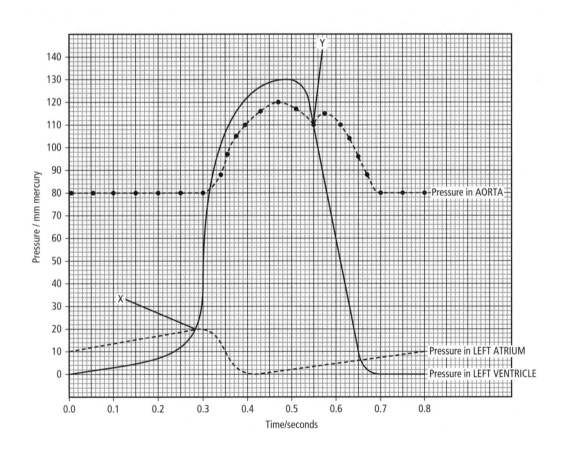

6. The diagram below shows an external view of the heart and its blood vessels.

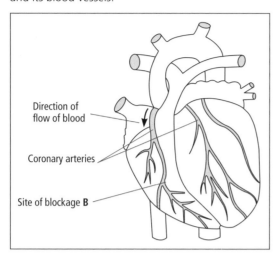

Direction of flow of blood

Coronary arteries

Site of blockage **B**

a. The coronary arteries supply heart tissue with useful substances. Coronary veins remove waste substances.
   i.   Name two useful substances the coronary arteries will supply. (2)
   ii   Name **one** waste substance the coronary veins will remove. (1)

b. The tissue forming the wall of the left ventricle responds when it is stimulated by electrical impulses.
   i.   Name this type of tissue. (1)
   ii. Describe how this tissue will respond when stimulated. (1)
   iii. Describe the effect of this response on the contents of the left ventricle. (2)
c. The coronary arteries can become blocked with a fatty deposit, leading to a heart attack.
   i.   State two likely causes of this type of blockage. (2)
   ii. A blockage occurs at point **B** in the coronary artery.
       On the diagram, shade in the parts of the artery affected by this blockage. (1)
d. Veins have different structures from arteries. State two features of veins and explain how these features enable them to function efficiently. (4)

CIE 0610 November '05 Paper 3 Q3

TRANSPORT IN ANIMALS: Crossword

ACROSS:
1 Lipid which may block blood vessels
3 Controls the rate of heartbeat – sometimes artificially
4 Tissue that contracts as the heart beats
7 Can be raised by stress of excitement
9 Vessels that carry blood away from the heart
10 Artery supplying the heart muscle with oxygen and glucose
12 With 2 down – prevents blood flowing back into heart
15 Strong tissue that prevents valves turning inside out
16 With 18 across – pumping chamber of the heart forcing blood to the body (4, 9)
18 See 16 across
19 Circulation for the lungs
21 With 20 down – receiving chamber for blood from the body
22 Vessels carrying blood back towards the heart

DOWN:
1 Tiny blood vessels for exchange of materials between blood and tissues
2 Structure that ensures one way blood flow
5 This can increase the risk of heart disease, as well as damaging the lungs
6 Main vein of the body
8 Benefits the heart, but too much can be dangerous
11 Blood pressure as heart relaxes
13 Blood pressure as heart contracts
14 The main artery
17 Operation to get past a blocked coronary artery
19 Count of the beats of the heart
20 See 21 across

Chapter 18:
Defence against disease

Natural defence systems of the body
prevent infection and disease

PATHOGENS CAUSE DISEASE. Many organisms may colonize the body of a human – the body is **warm, moist** and a **good food source**. These organisms may complete with human cells for nutrients or may produce by-products that are poisonous to human cells. This will affect the normal function of the cells, i.e it will cause 'disease'. There are three major groups of pathogens:

TYPE OF PATHOGEN	DISEASE	SYMPTOMS
VIRUS — Protein coat, Nucleic acid	Influenza	Fever (raised body temperature). Aching joints. Breathing problems. **Control by pain relief/rest/drinking fluids.**
BACTERIUM — Slime coat, Cell wall, 'Naked' DNA (not in chromosome)	Gonorrhoea	Painful to urinate – yellow discharge from penis or vagina. In the long term may cause blockage of sperm ducts/oviducts leading to sterility. **Control with antibiotics.**
FUNGUS — DNA in nucleus, Hypha secretes enzymes into food	Athlete's foot	Irritation to moist areas of skin (e.g. between toes). 'Cracked' skin may become infected. **Control with fungicide/drying powders.**

SKIN IS THE FIRST BARRIER: the outer layer of the skin, the **epidermis**, is waxy and impermeable to water and to pathogens (although many microorganisms can live on its surface). Where there are natural 'gaps' in the skin there may be protective secretions to prevent entry of pathogens, e.g.

ORIFICE	FUNCTION	PROTECTED BY
Mouth	Entry of food	Hydrochloric acid in stomach
Eyes	Entry of light	Lysozyme in tears
Ears	Entry of sound	Bacteriocidal ('bacteria-killing') wax

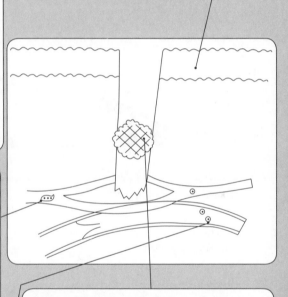

PHAGOCYTES DESTROY PATHOGENS BY INGESTING THEM

Phagocytes are large white blood cells. They are 'attracted to' wounds or sites of infection by chemical messages. They leave the blood vessels and destroy any pathogens they recognize.

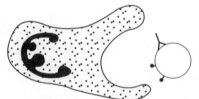

A pathogen is recognized by its surface proteins or by antibodies that 'label' it as dangerous.

Pathogens are destroyed by digestive enzymes secreted into 'food sac'.

LYMPHOCYTES PRODUCE PROTECTIVE ANTIBODIES

Lymphocytes are white blood cells that are found in the blood and in lymph nodes (swellings in the lymphatic system). They are stimulated by the presence of pathogens to manufacture and release special proteins called **antibodies**, which can recognize, bind to and help to destroy pathogens.

This end of the antibody acts as a 'signal to phagocytes to remove this pathogen

The forked end of the antibody 'recognizes' surface protein of pathogen

Surface protein identifies pathogen as non-human' cells

BLOOD CLOTTING PLUGS' WOUNDS: a natural defence, largely due to **blood proteins** and **platelets**. It is able to block any unnatural gaps in the skin.
• prevents excessive blood loss
• prevents entry of pathogens

DAMAGED PLATELETS

TORN CAPILLARIES

Inactive blood proteins e.g. FIBRINOGEN

ENZYMES

If CALCIUM IONS present

Mesh can trap red blood cells which dry out in form of a scab

Active blood proteins e.g. FIBRIN

FIBRES

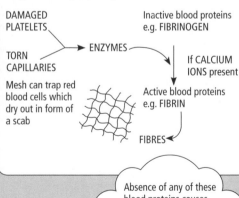

Absence of any of these blood proteins causes inefficient clotting → severe bleeding: **haemophilia.**

1. The diagram below shows the stages involved in one process of defence against disease.

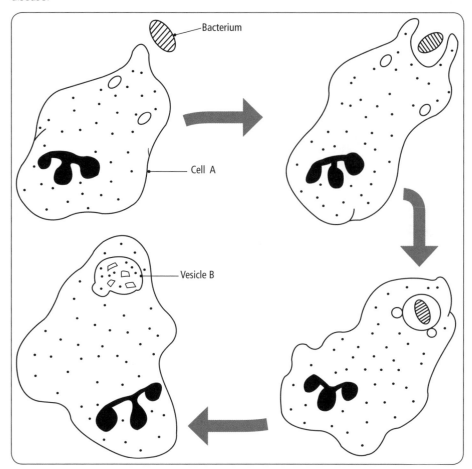

a. Name the cell labelled A. (1)

b. Name the process that the diagrams represent. (1)

c. How does cell A recognize the bacterium as a 'foreign' organism? (1)

d. What happens inside vesicle B to destroy the bacterium? (2)

e. Name two diseases in humans that are caused by bacteria (2)

f. What help can a doctor offer in the defence against bacterial diseases? (1)

2. The graph shows the levels of antibody in the blood following two injections of a vaccine.

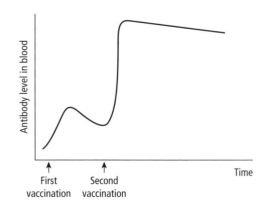

a. What is meant by the term 'vaccine'? (2)

b. What is an antibody? (1)

c. How does the response to the second injection differ from the response to the first injection? (3)

d. What is the advantage to a person in receiving the vaccine in two doses rather than as a single injection? (1)

e. This response of the body is an example of immunity. There are different ways of becoming immune, but each way is either 'active' or 'passive'. Complete the table below to show whether each of the methods described is active or passive. (5)

EXAMPLE	METHOD OF BECOMING IMMUNE	ACTIVE OR PASSIVE IMMUNITY
A	Receiving a vaccine	
B	Becoming infected with a virus	
C	Receiving a serum	
D	A baby feeding on colostrum	
E	A fetus receiving antibodies across the placenta	

3. a. The diseases AIDS, athlete's foot, malaria and cholera are caused by a number of different organisms.

  i. Match up each of these diseases with the type of organism that causes it.

A	AIDS	1	bacterium	B	cholera	2	protoctistan
C	athlete's foot	3	virus	D	malaria	4	fungus	(4)

  ii. Give *one* way in which each of the diseases is spread. (4)

 b. The effectiveness of different antibiotics can be investigated using cultures of bacteria grown on nutrient agar in petri dishes. The nutrient agar is heated to 120 °C in an autoclave, and then poured into sterile petri dishes and left until it hardens. The investigation, and the results, are shown in the diagram below:

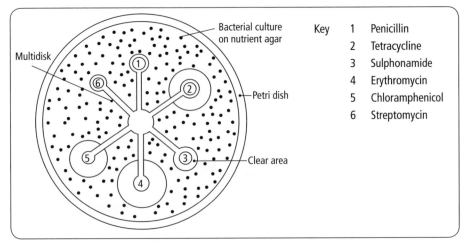

  i. The nutrient agar contains sugar, amino acids, vitamins and minerals. Why? (1)

  ii. Why is the nutrient agar pre-heated to a high temperature? (1)

  iii. Which is the most effective antibiotic among those used in this investigation? Explain how you arrived at your answer. (2)

  iv. Antibiotics can be **bacteriocidal** or **bacteriostatic**. Draw a graph to help you explain the meaning of these two terms. (4)

  v. Name two diseases that could not be treated with antibiotics. (2)

4. Polio is a serious disease caused by a virus. Polio can cause permanent paralysis. There are two methods of vaccinating against polio.

THE SALK METHOD	THE SABIN METHOD
Three injections of inactivated polio viruses are given during the first two years of life. Further 'booster' injections are given during childhood.	Inactive polio viruses are taken by mouth at about 6 months old. As a precaution, the dose may be repeated at around 14 years of age.

The graph shows the effect of vaccination on the amount of antibodies in the blood, using these two methods. The shaded area shows where there are an insufficient amount of antibodies for protection.

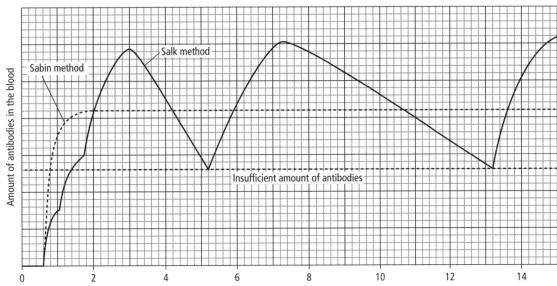

a. Suggest how the active form of the virus is treated to make it safe for use
   in a vaccine. (1)

b. Describe how the body responds to a vaccine, in order to become immune
   to polio. (3)

c. Use the information opposite to explain why the Sabin method of protection against polio is used more often than
   the Salk method.
   *To gain full marks in this question you should write your ideas in good English. Put them into a sensible order and
   use the correct scientific words.* (5)

d. i. Someone who has **not** been vaccinated is exposed to polio viruses.
      Neither the Salk nor the Sabin method would be an effective treatment for this person. Explain why. (1)

   ii. Suggest an alternative way of treating this person. (1)

5. Hepatitis B is a dangerous human disease. It is caused by a virus. The diagram shows how a vaccine against hepatitis B can
be made by genetic engineering.

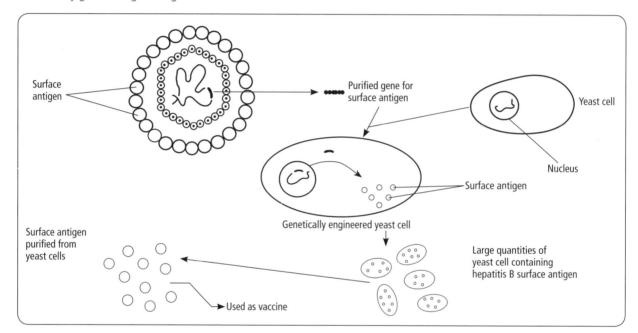

Use the diagram to answer the following questions.

a. i. What part of the virus is put into a yeast cell?

   ii. What part of the virus is produced by the yeast cell?

b. Before genetic engineering was developed, vaccines were made in a different way. The virus would be heated or
   treated chemically to stop it reproducing. Whole virus particles would be present in the vaccine.
      Explain why it would be safer to use a genetically engineered vaccine rather than to use a vaccine made directly
   from the hepatitis B virus. (2)

REVISION SUMMARY: Fill in the missing words

Complete the passage below, using terms from the list. You may use each term once, more than once or not at all.

DISEASE, PATHOGENS, FIBRINOGEN, PLATELETS, WHITE BLOOD CELLS, PROTEINS, BLOOD LOSS, RED BLOOD CELLS,
HAEMOPHILIA, FIBRIN

Damage to tissues results in the formation of clots. These have two basic functions – to reduce .... and to
prevent the entry of ...., which could cause ....
      Temporary clots form when the .... stick to each other, but more permanent wounds need stronger
barriers. The formation of more permanent clots involves several factors that are soluble .... in the blood. A
series of reactions eventually causes soluble .... to be converted to insoluble ...., which forms a
mesh. This mesh traps .... to form a strong barrier.
      The absence of a clotting factor may lead to a disease called ....

Chapter 19:

Breathing and gaseous exchange

Breathing, gaseous exchange and respiration

Oxygen gas is needed for respiration in humans. Carbon dioxide gas is a waste product of respiration. Breathing takes gases into and out of the body.

BREATHING: the action of drawing air into the body (inhaling) and pushing air and waste gases out (exhaling).

GASEOUS EXCHANGE: the movement of oxygen from inhaled air into the blood and the movement of carbon dioxide from the blood into the airways of the lungs.

Breathing in:
- muscles between the ribs contract, causing the chest to expand
- diaphragm flattens, increasing the volume of the chest and reducing gas pressure in the lungs
- air enters the lungs
- inhaled air has about 21 per cent **oxygen** and 0.03 per cent carbon dioxide

The airways of the lungs get smaller and smaller as they divide. The **bronchioles** end in thousands of tiny air sacs called **alveoli**. These have thin, moist walls so that gases can pass in and out easily.

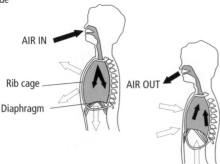

Breahing out:
- muscles between the ribs relax
- diaphragm raised up, decreasing the volume of the chest cavity and increasing gas pressure in the lungs
- air 'pushed' out of the lungs
- exhaled air has about 4 per cent **carbon dioxide**, 17 per cent oxygen, and is saturated with water vapour

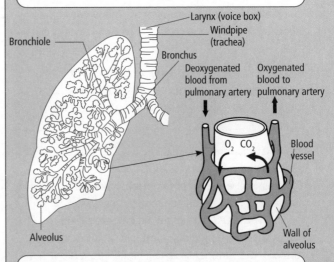

Each **alveolus** is covered by thin capillaries. Oxygen diffuses into the blood where it is taken up by red blood cells. Carbon dioxide diffuses from the blood into the alveolus, ready to be exhaled.

RESPIRATION IN HUMANS: We usually respire aerobically, using oxygen to release energy from our food.

This carbon dioxide can be detected because it turns **lime water** milky: the carbon dioxide reacts with the calcium ions in the lime water to form small particles of chalk.

AEROBIC RESPIRATION

glucose + oxygen ⟶ carbon dioxide + water + [energy]

ANAEROBIC RESPIRATION

When the cells have no oxygen available they can respire **anaerobically**.

glucose ⟶ lactic acid + [energy]

In plants, the waste product of anaerobic respiration is alcohol. In animals it is **lactic acid**.

EXERCISE AND BREATHING

- Exercise produces more **carbon dioxide**
- Extra carbon dioxide **lowers the pH of the blood**
- Change in pH **stimulates the breathing control centre in the brain**
- Breathing control centre **increases rate and depth of breathing**
- Excess carbon dioxide is cleared from the blood

PROTECTING THE AIRWAYS

After exercise oxygen is needed for recovery!

Cilia – fine 'hairs' on surface of cell. These can beat in a coordinated way to carry mucus (with trapped pathogens and particles) away from the lung surfaces.

Nucleus

Basement membrane – holds the cells in place

Goblet cell – produces sticky mucus and releases it onto the surface of the cells.

Mucus

During sport or strenuous activity, muscle cells can run out of oxygen. They then start to respire anaerobically. The lactic acid produced builds up in the muscles and can cause painful 'cramps' until you have taken in enough oxygen to break down the lactic acid and repay the **oxygen debt**.

1. Match up the two lists below to explain the ideal features of a gas exchange surface.

	FEATURE		VALUE IN GAS EXCHANGE
A	moist	a	gases do not have to diffuse very far
B	large surface area	b	oxygen is quickly removed so that diffusion can continue
C	thin	c	increase the number of molecules that can diffuse across at the same time
D	well ventilated	d	cells die if they dry out
E	good blood supply	e	regular supply of fresh air keeps up the concentration gradients for oxygen and carbon dioxide

(5)

2. Complete the following paragraphs. The words in the following list may be used once, more than once or not at all.

GILLS, OXYGEN, ENERGY, THIN, AMOEBA, NITROGEN, MOIST, SPIRACLES, RESPIRATION, EXCRETION, SURFACE AREA, BLOOD, TRACHEOLES, VENTILATION, CARBON DIOXIDE

All living organisms require........................., which is released from the process of ....
The most efficient form of this process requires the gas ....and produces the waste gas........................To keep this energy – releasing process going the organism must have a gas exchange surface – this surface has certain properties, it has a large........................, a........................ membrane so that diffusion distances are short and a........................ layer (since cells die if they dry out). In addition the most advanced systems have a means of........................ to move the gases over the surface, and are close to a........................supply to transport gases between the surface and the living tissues.

In simple organisms such as........................enough oxygen can diffuse through the outer cell membrane, but larger organisms require more oxygen than diffusion alone can supply. Insects have small 'holes', called........................, in their body covering – these connect to tubes called........................ that lead directly to the working tissues. More active and larger animals may have a specialized gas exchange surface, such as the........................found in fish. (13)

3. a. The diagram shows the position of the diaphragm and ribs at rest and after inhaling.

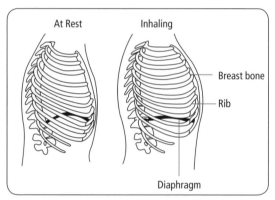

At Rest   Inhaling

Breast bone
Rib
Diaphragm

Explain how movements of the diaphragm and ribs cause air to enter the lungs during inhalation. (4)

b. A scientist investigated the effect of increasing carbon dioxide concentration on the breathing rate and on the total volume of air breathed per minute. The results are shown in the graph below.

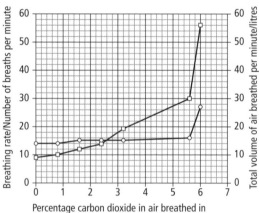

Percentage carbon dioxide in air breathed in

-○- Breathing rate
-□- Total volume of air breathed per minute/litres

i. Use the graph to describe the effect of increasing the carbon dioxide concentration on the total volume of air breathed per minute. (2)
ii. Suggest why the total volume of inhaled air is **not** directly proportional to the rate of breathing. (2)

4. The tables below contain information on the effects of smoking on health.

CIGARETTES SMOKED PER DAY	ANNUAL DEATH RATE PER 1000 MEN
0	0.1
5	0.4
10	0.6
15	1.0
20	1.5
25	1.7
30	2.0
35	2.4
40	2.7
45	3.0
50	3.6

PERIOD SINCE GIVING UP SMOKING / YEARS	ANNUAL DEATH RATE PER 1000 MEN
0	1.80
5	0.96
10	0.55
15	0.31
20	0.25
25	0.20

a. Plot these figures as two graphs. (8)
b. Out of a group of 5000 men smoking 20 cigarettes per day, how many are likely to die of lung cancer in one year? (1)
c. How long does it take to reduce the risk of lung cancer by 60 per cent by giving up smoking? Give your answer to the nearest year. (1)
d. Name the addictive substance found in tobacco smoke. (1)
e. Smokers often have blood that has lost some of its 'redness': this can be seen by the lips taking on a pale blue colour. Suggest a reason for this. (1)
f. Name two other diseases that can be caused by cigarette smoking. (2)

5. A student's breathing was monitored before and after vigorous exercise. The student breathed in and out through special apparatus. The graphs show the changes in the volume of air inside the apparatus.
Each time the student breathed in, the line on the graph dropped. Each time the student breathed out, the line went up.

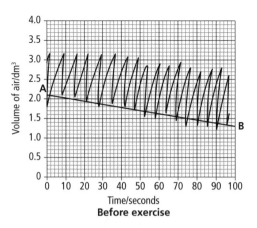
Before exercise

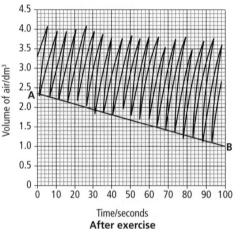
After exercise

a. How many times did the student breathe in per minute:
    before exercise;
    after exercise? (1)
b. On each graph, the line **A–B** shows how much oxygen was used. The rate of oxygen use before exercise was 0.5 dm³ per minute.
    Calculate the rate of oxygen use after exercise.

    Rate of oxygen use after exercise =
    ....dm³ per minute (2)

c. The breathing rate and the amount of oxygen used were still higher after exercise, even though the student sat down to rest.
    Why were they still higher? (4)

6. The figure below shows an alveolus in which gaseous exchange takes place.

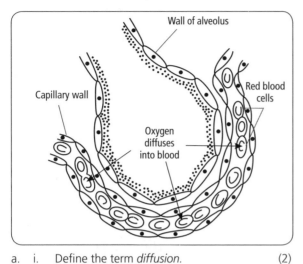

Wall of alveolus

Capillary wall

Red blood cells

Oxygen diffuses into blood

a. i.   Define the term *diffusion*.   (2)
    ii. State what causes oxygen to diffuse into the blood from the alveoli.   (1)
    iii. List three features of gaseous exchange surfaces in animals, such as humans.   (3)

b. i.   At high altitudes there is less oxygen in the air than at sea level. Suggest how this might affect the uptake of oxygen in the alveoli.   (2)
    ii. In the past some athletes have cheated by injecting themselves with extra red blood cells before a major competition.
         Predict how this increase in red blood cells might affect their performance.   (2)

CIE 0610 November '06 Paper 2 Q9

Revision Summary: Fill in the missing words

Complete the following paragraph by filling in the missing words. The words in the following list may be used once, more than once or not at all.

ENERGY, HYDROGENCARBONATE, LEFT ATRIUM, PULMONARY, RESPIRATION, THIN. OXYGEN, SURFACE AREA, RENAL, ALVEOLI, CAPILLARIES, ARTERIES, TRACHEA, RIGHT VENTRICLE.

Deoxygenated blood arrives at the lungs in the .... arteries. Oxygen has been removed from the blood by cells that are carrying out .... to release .... needed to carry out their functions. This blood also contains a relatively high concentration of the gas........................, which is carried dissolved in the plasma as .... ions.
Each artery branches many times to form........................ which are well adapted to allow the exchange of gases because they are........................-walled and have a very large........................ These small vessels lie very close to the .... of the lungs, and it is here that gas exchange takes place. The gas ....moves out of the blood and the gas ....moves into the blood. Both gases move by the process of .... Oxygenated blood then leaves the lungs in the........................vein that returns blood to the heart at the chamber called the ....   (14)

Respiration may be aerobic or anaerobic

Respiration is the chemical process in which food, usually glucose, is oxidized to release energy. The energy is used to carry out the work needed to keep organisms alive:
- mechanical work (movement of muscles),
- growth and repair (by cell division)
- chemical work (active transport and chemical building in plants and animals)
- production of heat needed for **metabolic** (life) processes, e.g. providing the optimum temperature for the action of enzymes

ATP (adenosine triphosphate) is the energy compound used in cells.

ATP

ENERGY from FOOD

ENERGY for WORK

ADP + Phosphate

AEROBIC RESPIRATION

In most organisms, respiration uses **oxygen**. This is called **aerobic respiration**, and takes place in the MITOCHONDRIA.

The chemical reactions are complex but we can summarize aerobic respiration in this equation:

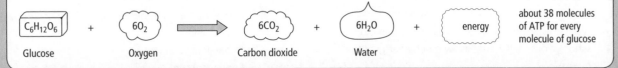

$C_6H_{12}O_6$ + $6O_2$ $\Longrightarrow$ $6CO_2$ + $6H_2O$ + energy   about 38 molecules of ATP for every molecule of glucose

Glucose   Oxygen   Carbon dioxide   Water

ANAEROBIC RESPIRATION

In some cases, respiration takes place **without** oxygen. This **anaerobic respiration**
- only partly breaks down the glucose
- releases less energy
- takes place in the cytoplasm

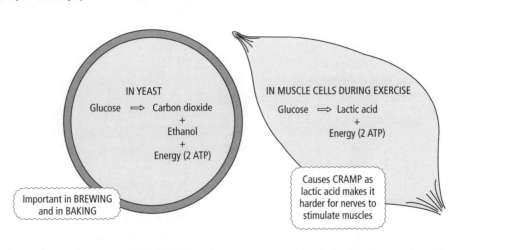

IN YEAST

Glucose $\Longrightarrow$ Carbon dioxide
+
Ethanol
+
Energy (2 ATP)

IN MUSCLE CELLS DURING EXERCISE

Glucose $\Longrightarrow$ Lactic acid
+
Energy (2 ATP)

Important in BREWING and in BAKING

Causes CRAMP as lactic acid makes it harder for nerves to stimulate muscles

1. Respiration is a feature of living organisms. Make a list of five other characteristics of living organisms. (5)

2. The diagram shows how much energy is required for various activities. The figures are given in kilojoules (kJ) per hour.

Sleeping : 300   Standing 450   Walking 1000

School work : 650   Running 2200   Swimming 3200

   a. Plot this information on a bar chart. (4)
   b. During a typical school day Sam stood for 1 hour, walked for 1 hour, ran for half an hour and did school work for 4 hours. How much energy did he use? (2)
   c. Sleeping uses 300 kJ in one hour. Why is energy needed during periods of sleeping? (2)
   d. Suggest why swimming uses more energy than running. (2)

3. a. A student investigated the energy content of a seed.

   A seed was weighed and its mass recorded in the table opposite. The seed was firmly attached to the end of a mounted needle. A large test tube containing 20 cm³ of water was held in a clamp stand, with a thermometer and a stirrer. The apparatus is shown below.

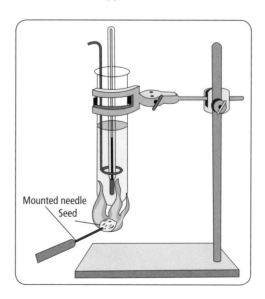

Mounted needle
Seed

- The temperature of the water at the start was recorded.
- The seed was set alight by placing it in a flame for a few seconds.
- The burning seed was held under the test tube until the seed was completely burnt.
- The water was stirred immediately. The highest temperature of the water was recorded.

   i. Complete the table by calculating the rise in temperature. (1)

MASS OF SEED / g	0.5
VOLUME OF WATER / cm³	20
TEMPERATURE AT THE START / °C	29
HIGHEST TEMPERATURE / °C	79
RISE IN TEMPERATURE / °C	

The energy contained in the seed can be calculated using the formula below.

$$\text{energy} = \frac{\text{volume of water} \times \text{rise in temperature} \times 4.2}{\text{mass of seed} \times 1000}$$

   ii. Using the formula calculate the energy content of the seed.
   Show your working.
   Energy content ....kJg⁻¹ (2)

The same method was used to find the energy content of some food substances. The results are shown in the next table.

FOOD SUBSTANCE	Starch	Sugar	Fat	Protein
MASS OF FOOD BURNT / g	0.62	0.54	0.56	0.40
STARTING TEMPERATURE / °C	31	30	30	31
FINAL TEMPERATURE / °C	65	59	90	52
RISE IN TEMPERATURE / °C	34	29	60	21
ENERGY CONTENT / kJg⁻¹	4.61	4.51	9.00	4.41

   iii. Plot a suitable graph to compare the energy content per gram of the four different food substances and the seed from (a)(ii). (4)
   iv. Use this information to suggest the main food substance present in the seed. (1)
   b. Describe how you would test for the presence of reducing sugars in a seed. (3)

CIE 0610 June '06 Paper 6 Q1

4. The diagrams below show four stages of an investigation in which two mice were each fed with glucose solution.

One of the mice was fed with a glucose solution containing radioactive carbon (^{14}C). The other mouse was fed with a normal (non-radioactive) glucose solution.

Mice fed on glucose solution containing low levels of radioactive carbon (^{14}C) for short periods of time suffer no ill effects.

Both mice were approximately the same size.

a. i.   Name the gas which was responsible for
         turning the limewater milky at stage 2.   (1)
    ii. Name the living process in the cells which
         produces this gas.
    iii. This process releases energy. State **three** uses
         of this energy in mice.   (3)

b. At stage 2 the air entering the bell jar containing the mouse contained no carbon dioxide. Describe a method by which the carbon dioxide could have been removed.   (2)
c. Explain the importance of the mice being approximately the same size.   (2)
d. Describe the sequence of processes that lead to filter paper A becoming radioactive.   (5)

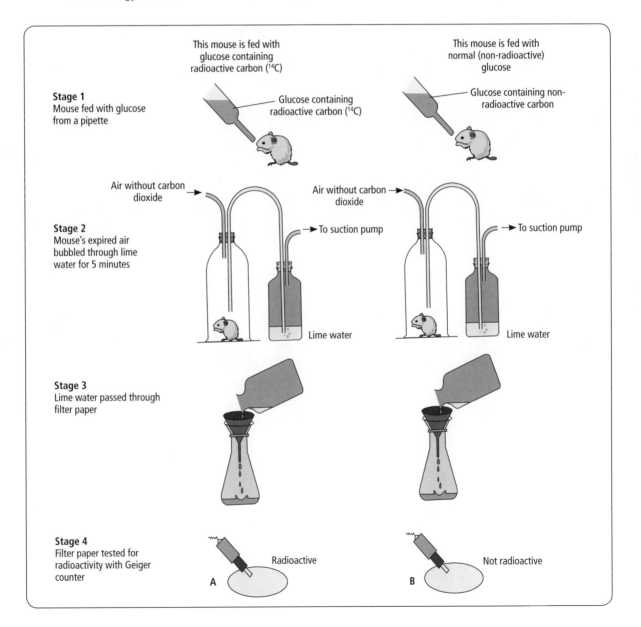

This mouse is fed with glucose containing radioactive carbon (^{14}C)

This mouse is fed with normal (non-radioactive) glucose

Stage 1
Mouse fed with glucose from a pipette

Glucose containing radioactive carbon (^{14}C)

Glucose containing non-radioactive carbon

Stage 2
Mouse's expired air bubbled through lime water for 5 minutes

Air without carbon dioxide

To suction pump

Air without carbon dioxide

To suction pump

Lime water

Lime water

Stage 3
Lime water passed through filter paper

Stage 4
Filter paper tested for radioactivity with Geiger counter

Radioactive

A

Not radioactive

B

5. Here are a number of statements about aerobic respiration. State whether each of them is TRUE or FALSE.

Aerobic respiration

a. uses water

b. always requires oxygen

c. occurs only in animal cells

d. releases energy

e. produces carbon dioxide

f. is controlled by enzymes

g. must have sugar as a starting material

h. releases heat

i. releases water

j. is affected by temperature

k. occurs in all living cells   (10 × 1)

REVISION SUMMARY: Fill In the missing words

Complete the following paragraph. Use terms from the list below – you may use each term once, more than once or not at all.

ANAEROBIC, MOIST, LIVING CELLS, GROWTH, ANIMALS, GLUCOSE, AEROBIC, OXYGEN, CARBON DIOXIDE, HEAT, MOVEMENT, WATER, ENERGY.

Respiration is a process that occurs in all .... The raw materials are .... and, sometimes .... The purpose of the process is to release ...., and it is more efficient under .... conditions. One product of respiration can be used to carry out work: this work includes .... and ...., but because respiration is never 100 per cent efficient .... is also released and may be used to keep up body temperature in mammals and birds.   (8)

Chapter 21:
Excretion and osmoregulation

Kidney structure and function

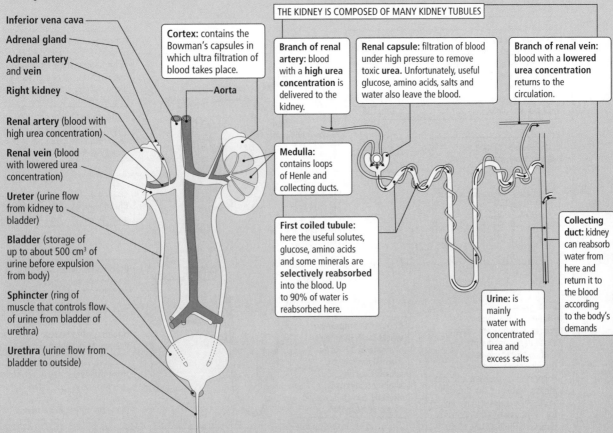

Inferior vena cava

Adrenal gland

Adrenal artery and vein

Right kidney

Renal artery (blood with high urea concentration)

Renal vein (blood with lowered urea concentration)

Ureter (urine flow from kidney to bladder)

Bladder (storage of up to about 500 cm³ of urine before expulsion from body)

Sphincter (ring of muscle that controls flow of urine from bladder of urethra)

Urethra (urine flow from bladder to outside)

Cortex: contains the Bowman's capsules in which ultra filtration of blood takes place.

Aorta

THE KIDNEY IS COMPOSED OF MANY KIDNEY TUBULES

Branch of renal artery: blood with a **high urea concentration** is delivered to the kidney.

Renal capsule: filtration of blood under high pressure to remove toxic **urea**. Unfortunately, useful glucose, amino acids, salts and water also leave the blood.

Branch of renal vein: blood with a **lowered urea concentration** returns to the circulation.

Medulla: contains loops of Henle and collecting ducts.

First coiled tubule: here the useful solutes, glucose, amino acids and some minerals are **selectively reabsorbed** into the blood. Up to 90% of water is reabsorbed here.

Collecting duct: kidney can reabsorb water from here and return it to the blood according to the body's demands

Urine: is mainly water with concentrated urea and excess salts

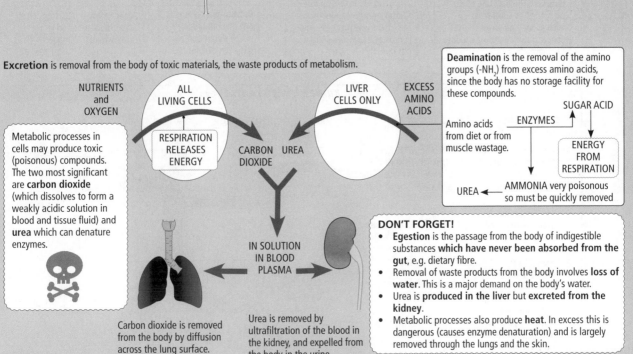

Excretion is removal from the body of toxic materials, the waste products of metabolism.

Deamination is the removal of the amino groups (-NH₂) from excess amino acids, since the body has no storage facility for these compounds.

NUTRIENTS and OXYGEN

ALL LIVING CELLS

RESPIRATION RELEASES ENERGY

CARBON DIOXIDE

UREA

LIVER CELLS ONLY

EXCESS AMINO ACIDS

Amino acids from diet or from muscle wastage.

ENZYMES

SUGAR ACID

ENERGY FROM RESPIRATION

AMMONIA very poisonous so must be quickly removed

UREA

Metabolic processes in cells may produce toxic (poisonous) compounds. The two most significant are **carbon dioxide** (which dissolves to form a weakly acidic solution in blood and tissue fluid) and **urea** which can denature enzymes.

IN SOLUTION IN BLOOD PLASMA

Carbon dioxide is removed from the body by diffusion across the lung surface.

Urea is removed by ultrafiltration of the blood in the kidney, and expelled from the body in the urine.

DON'T FORGET!
- **Egestion** is the passage from the body of indigestible substances **which have never been absorbed from the gut**, e.g. dietary fibre.
- Removal of waste products from the body involves **loss of water**. This is a major demand on the body's water.
- Urea is **produced in the liver** but **excreted from the kidney**.
- Metabolic processes also produce **heat**. In excess this is dangerous (causes enzyme denaturation) and is largely removed through the lungs and the skin.

Kidney failure

If one or both kidneys fail then dialysis is used or a transplant performed to keep urea and solute concentration in the blood constant.

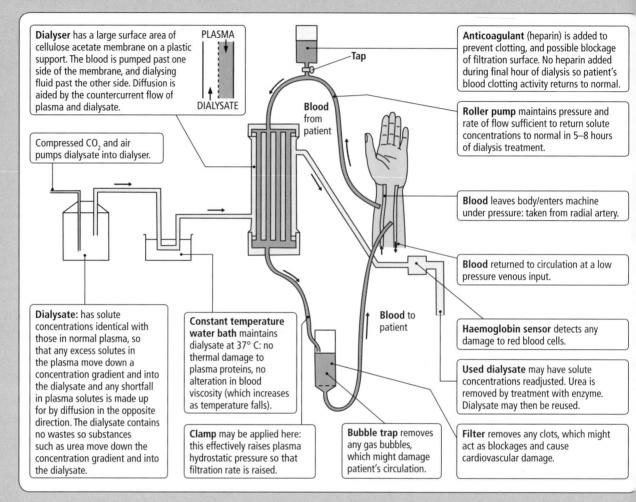

Dialyser has a large surface area of cellulose acetate membrane on a plastic support. The blood is pumped past one side of the membrane, and dialysing fluid past the other side. Diffusion is aided by the countercurrent flow of plasma and dialysate.

PLASMA

DIALYSATE

Compressed CO_2 and air pumps dialysate into dialyser.

Dialysate: has solute concentrations identical with those in normal plasma, so that any excess solutes in the plasma move down a concentration gradient and into the dialysate and any shortfall in plasma solutes is made up for by diffusion in the opposite direction. The dialysate contains no wastes so substances such as urea move down the concentration gradient and into the dialysate.

Constant temperature water bath maintains dialysate at 37° C: no thermal damage to plasma proteins, no alteration in blood viscosity (which increases as temperature falls).

Clamp may be applied here: this effectively raises plasma hydrostatic pressure so that filtration rate is raised.

Tap

Blood from patient

Blood to patient

Bubble trap removes any gas bubbles, which might damage patient's circulation.

Anticoagulant (heparin) is added to prevent clotting, and possible blockage of filtration surface. No heparin added during final hour of dialysis so patient's blood clotting activity returns to normal.

Roller pump maintains pressure and rate of flow sufficient to return solute concentrations to normal in 5–8 hours of dialysis treatment.

Blood leaves body/enters machine under pressure: taken from radial artery.

Blood returned to circulation at a low pressure venous input.

Haemoglobin sensor detects any damage to red blood cells.

Used dialysate may have solute concentrations readjusted. Urea is removed by treatment with enzyme. Dialysate may then be reused.

Filter removes any clots, which might act as blockages and cause cardiovascular damage.

KIDNEY TRANSPLANTATION

Transplantation may be necessary as renal dialysis is inconvenient for the patient and costly.

Kidney transplants have a high success rate because:
1. The vascular connections are simple.
2. Live donors may be used, so very close blood group matching is possible.
3. Because of 2 there are fewer immunosuppression-related problems in which the body's immune system reacts against the new kidney.

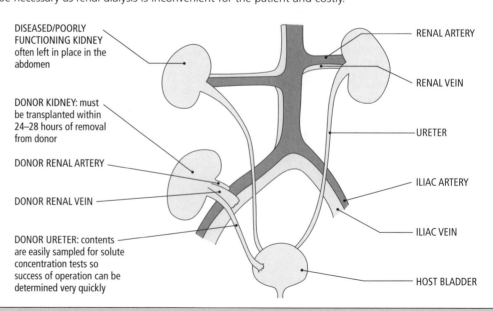

DISEASED/POORLY FUNCTIONING KIDNEY often left in place in the abdomen

DONOR KIDNEY: must be transplanted within 24–28 hours of removal from donor

DONOR RENAL ARTERY

DONOR RENAL VEIN

DONOR URETER: contents are easily sampled for solute concentration tests so success of operation can be determined very quickly

RENAL ARTERY

RENAL VEIN

URETER

ILIAC ARTERY

ILIAC VEIN

HOST BLADDER

1. The diagram shows water loss and uptake for an adult human.

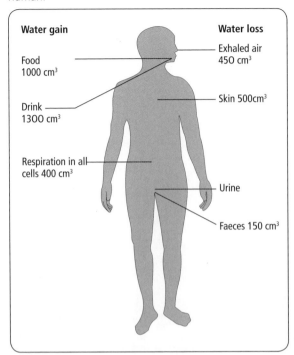

Water gain

Food
1000 cm³

Drink
1300 cm³

Respiration in all
cells 400 cm³

Water loss

Exhaled air
450 cm³

Skin 500cm³

Urine

Faeces 150 cm³

The kidneys keep the body's water content at its optimum level by controlling the water lost in the urine.

a. Use information from the diagram to calculate the mean daily loss of water in urine. Show your working. (2)

b. Describe how the kidneys control the water content of the body. Use the term 'negative feedback' in your answer. (4)

c. Sometimes kidneys fail, and the composition of the blood may change to a dangerous extent. The condition may be treated by dialysis or with a kidney transplant.

   i. Give two possible advantages of using a dialysis machine rather than having a kidney transplant

   ii. Give two possible disadvantages of using a dialysis machine rather than having a kidney transplant. (4)

2. The diagram shows the process of filtration in the kidney.

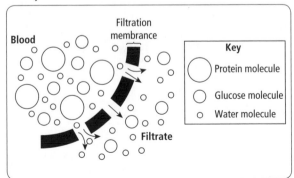

Filtration
membrane

Blood

Key

Protein molecule

Glucose molecule

Water molecule

Filtrate

a. Use information in the diagram and your own knowledge of how the kidney works to explain why:

   i. protein molecules are not normally present in urine; (1)

   ii. glucose molecules are not normally present in urine. (3)

b. An athlete trained for two hours on a hot summer's day. At the end of the training session, the athlete had a higher concentration of antidiuretic hormone (ADH) in his blood than at the start of the training session. Explain why. (4)

3. a. Define the terms

   i. excretion, (1)

   ii. egestion. (1)

b. The kidney is an excretory organ.
It produces urine that contains urea.

   i. State where in the body urea is formed. (1)

   ii. State what urea is formed from. (1)

c. This diagram shows the urinary system and its blood supply.

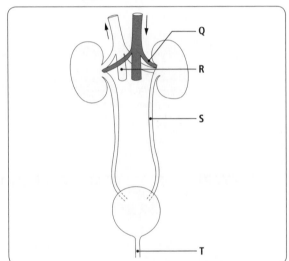

Q

R

S

T

Name the parts labelled **Q**, **R**, **S** and **T**. (4)

d. Complete the table below to show which components of the blood are also part of the urine of a healthy person.
Use ticks (√) and crosses (X). Two boxes have already been completed. (2)

COMPONENT OF BLOOD	PRESENT IN URINE
Glucose	
Red blood cells	
Salts	
Urea	√
Water	
White blood cells	X

CIE 0610 November '05 Paper 2 Q3

4. A patient has kidney failure. Every few days the patient goes into hospital for dialysis treatment.
The graph shows the concentration of urea in the patient's blood over a period of two weeks.
The concentration of urea in a healthy person's blood is shown for comparison.

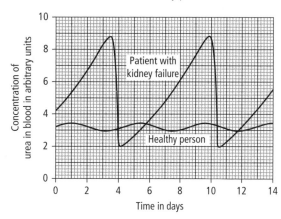

a. Suggest why the concentration of urea in the patient's blood rises between
   dialysis treatments. (1)

b. If the concentration of urea in the blood rises to 9 units, the patient may die.
   The dialysis machine reduces the concentration of urea in the patient's blood
   to below the concentration in the blood of a healthy person.
   Suggest **one** advantage of this to the patient. (1)

c. One alternative to treatment with a dialysis machine is to have a kidney transplant.
   The donor kidney is of similar tissue type to that of the patient.
   However, the transplanted kidney may still be rejected by the patient's immune system.
   Slate **two** other treatments which may help to prevent rejection of the transplanted kidney.
   Explain your answers. (4)

EXCRETION AND OSMOREGULATION: Crossword

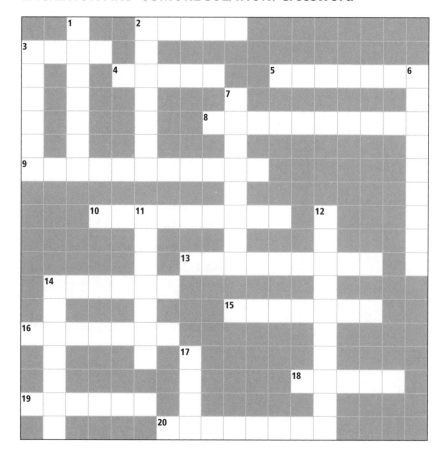

ACROSS:
2 Expelled from the body to remove urea
3 Nitrogen-containing waste product
4 This loop makes conditions which control water conservation
5 Storage sac for 3 across
8 Process that produces fluid in the kidney tubule
9 Delivers blood to the kidney (5, 6)
10 Ring of muscle that prevents embarrassing urination
13 The removal of waste materials from the body
14 Liquid part of the blood
15 Important source of energy reclaimed from fluid in the kidney tubule
16 Capsule where 8 across goes on
18 Urea is produced here
19 Urea is removed from the blood here
20 Treatment for a failing kidney

DOWN:
1 Kidney tubule
2 Tube from bladder to the environment
3 Tube from kidney to bladder
6 Carries 'purified' blood away from the kidney
7 With 11 down and 17 down: controls water conservation by the kidney
11 See 7 down
12 Deamination helps to remove these compounds if they are in excess of the body's needs
14 Solute that should never leave the plasma
17 See 7 down

Chapter 22:

Hormones and the endocrine system

The endocrine system

The endocrine system uses chemical 'messengers' called **hormones** to co-ordinate the response of organs to stimuli and changes in the environment.

The **endocrine system** is made up of **glands** that can release special chemicals called hormones into the blood stream. Detectors on body organs, e.g. the heart, detect changes in the level of hormones in the blood. When the hormone level changes, the organ responds.

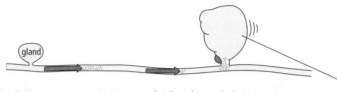

1 Endocrine gland responds to change in environment
2 Gland releases hormone
3 Hormone distributed throughout the body
4 Detector on organ senses change in hormone level
5 Organ responds, e.g. heart beats faster

Glands, hormones and their functions

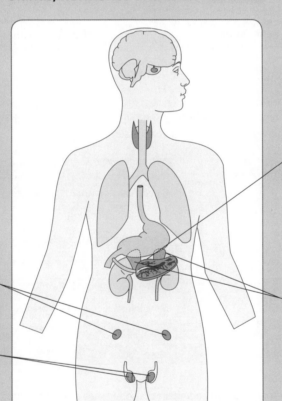

Pancreas: produces insulin, which controls the level of sugar in the blood.

Diabetes is a disease where the body cannot control the blood sugar level properly. Diabetes sufferers have to control their diet carefully. Some have to inject themselves with insulin.

Ovaries: in females the ovaries produce **oestrogen** and **progesterone**. These cause changes at puberty and control the menstrual cycle.

Adrenal glands: produce adrenalin where there is danger. This prepares the body for fighting or running away (**fight** or **flight** response) for example:
• increases blood glucose concentration
• increases heart rate
• increases breathing rate

Testes: in males, the testes produce **testosterone** which causes changes at pubery.

Comparing the nervous system with the endocrine system

	WHAT DETECTS CHANGE? (RECEPTOR)	WHAT IS THE 'MESSENGER'?	HOW IS THE MESSAGE CARRIED?	WHAT RESPONDS? (EFFECTOR)
NERVOUS SYSTEM	sensory receptors	electrical impulse	nerves – very fast	a muscle or a cell
ENDOCRINE SYSTEM	gland (via brain and bloodstream)	chemical (hormone)	bloodstream – slower	one or more organs

1. Bovine somatotrophin (BST) is a natural growth hormone. The hormone can also be produced by genetic engineering, and is valuable because it can affect meat and milk production in cattle. The hormone can be injected into the bloodstream of cattle, or supplied in a gelatin coat so that it can be given in cattle food.

   The bar chart shows the results of some experiments carried out at a cattle research institute.

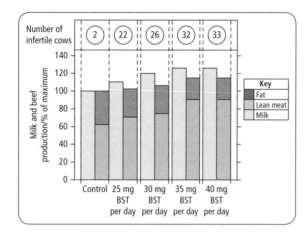

   a. What are the advantages to a farmer of treating cattle with BST? (1)

   b. Suggest a possible disadvantage to the farmer of treating cattle with BST. (1)

   c. Suggest a possible disadvantage to the human consumers of beef treated with BST. (2)

   d. Explain why enclosing the BST in a gelatin coat is necessary if the hormone is to be given in the cattle food.

   e. i. Chickens are also treated with growth hormones. One such hormone is the female sex hormone, oestrogen. This hormone makes chickens grow more quickly and retain more water in their tissues, so they become heavier in a shorter time.

      ii. Male body builders require a high protein diet. Many find that beef is too expensive for regular consumption, and so eat chicken and chicken products as often as twice a day. Body builders often report 'feminization', i.e. the growth of breasts, the shrinkage of their testes and the loss of facial hair.

   Explain how these two observations could be linked. (2)

2. One important hormone in mammals is **adrenaline**. This has many effects on body systems. Match the columns below to show the benefits of adrenaline secretion.

	EFFECT OF ADRENALINE		BENEFIT TO BODY
A	Pupil of eye widens	1	Blood moves more quickly to working tissues
B	Heart beats faster and deeper	2	Excess heat can be lost
C	Glycogen is hydrolysed	3	More oxygen enters the blood
D	More blood delivered to brain	4	More oxygen and glucose are delivered to respiring muscle cells
E	More blood delivered to skin	5	Allows more light onto the retina
F	Less blood delivered to the gut	6	More information can be processed and actions taken
G	Breathing rate increase	7	Digestion will be slowed
H	Blood flow to muscle is increased	8	More glucose is available for respiration

(7)

Use information from this table to explain why adrenaline is sometimes called the 'flight, fright or fight' hormone. (3)

3. The normal blood glucose level is 1 mg per cm³. Ten people with normal blood glucose levels were tested for blood glucose and plasma insulin levels over a period of six hours. The mean values for these measurements were calculated and recorded. The test period included two meals and a session of exercise. The results are shown in the table below.

TIME/ HOURS	ACTIVITY	BLOOD GLUCOSE LEVEL/mg cm⁻³	PLASMA INSULIN LEVEL/μg cm⁻³
0	Meal eaten	1.0	10
0.5		1.5	20
1.0		1.0	40
1.5		0.8	25
2.0	Exercise started	0.8	15
2.5	Exercise finished	1.2	10
3.0		1.0	20
3.5		1.0	10
4.0	Meal eaten	1.0	10
4.5		1.4	20
5.0		1.0	35
5.5		0.8	40
6.0		0.8	10

a. Present all the data in the form of a graph. (6)
b. What effect does the period of exercise have on the blood glucose and plasma insulin levels? Explain your answer. (3)
c. Suggest two other hormones that would change in concentration in the blood during the period of exercise. Why are these hormones important? (4)
d. Why is it good experimental technique to
  i. take **mean** values for blood glucose and plasma insulin levels
  ii. use only subjects with normal glucose levels? (2)
e. How long after a meal does it take for the blood glucose level to return to normal? (1)

4. Complete the table with the names of the three missing hormones. Choose your answers from the following list.
adrenaline
auxin
follicle-stimulating hormone
insulin
oestrogen
progesterone
testosterone

Hormone	What it does in the human body
	lowers the amount of sugar in the blood
	causes development of the female secondary sexual characteristics
	causes development of the milk-secreting glands in the breasts during pregnancy

(3)

5. Insulin is a hormone produced to control blood glucose levels. Diabetics do not have a natural ability to control these levels.
a. Define the term *hormone*. (2)
b. With reference to the pancreas and the liver, describe the role of insulin in controlling blood glucose levels. (4)
c. • Insulin is a protein.
   • Diabetics can control their blood glucose levels artificially by injecting insulin.
   • Many medicines are swallowed as tablets.
   Explain what would happen to the insulin in the stomach if it was swallowed as a tablet. (2)

CIE 0610 June '06 Paper 3 Q4

6. Read the following passage which is from an advice book for diabetics.

Insulin Reactions

Hypoglycaemia or 'hypo' for short, occurs when there is too little sugar in the blood. It is important always to carry some form of sugar with you and take it immediately you feel a 'hypo' start. A hypo may start because:
• you have taken too much insulin, or
• you are late for a meal, have missed a meal altogether, have eaten too little at a meal, or
• you have taken a lot more exercise than usual.

The remedy is to take some sugar.

An insulin reaction usually happens quickly and the symptoms vary – sweating, treambling, tingling of the lips, palpitations, hunger, pallor, blurring of the vision, slurring of speech, irritability, difficulty in concentration. Do not wait to see if it will pass off, as an untreated 'hypo could lead to unconsciousness.

a. Many diabetics need to take insulin.
   i. Explain why. (2)
   ii. Explain why there is too little sugar in the blood if too much insulin is taken. (3)
   iii. Explain why there is too little sugar in the blood if the person exercises more than usual. (3)
b. Suggest why sugar is recommended for a 'hypo', rather than a starchy food. (3)
c. Explain how the body of a healthy person restores blood sugar level if the level drops too low. (3)
d. Explain, using insulin as an example, what is meant by negative feedback. (3)

REVISION SUMMARY: Fill in the gaps

Complete the paragraph by filling in the gaps with words from the list below. You may use each word once, more than once or not at all.

ADRENALINE, HORMONE, BLOODSTREAM, DIGESTIVE SYSTEM, ENDOCRINE ORGAN, INSULIN, TARGET ORGAN, TRACHEA, OESOPHAGUS

In all mammals there is a chemical coordination system. This system uses chemical messengers called ...., which are secreted by the ...., travel in the .... and have an effect on a .... An example of this type of chemical messenger is ...., which, during periods of fear or anxiety, relaxes the ...., allowing the person to breathe more easily. (6)

Neurones (nerve cells) carry information through the nervous system

Dendrites 'collect' information from other cells.

Cell body controls the metabolism of the nerve cell.

Axon is extended to carry information over long distances.

Fatty sheath (myelin) is electrical insulation which prevents distortion of the nervous impulse by activities of neighbouring cells.

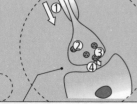

DIRECTION OF IMPULSE

Node of Ranvier allows rapid conduction of impulses by forcing message of 'jump' from one node to the next.

End plate 'synapses' with another nerve cell, a muscle or a gland.

Drugs may interfere with synaptic activity e.g. Amphetamine ('Speed') stimulates nervous activity because its chemical structure is similar to the neurotransmitter and it mimics it in the synaptic gap.

The synapse is the junction between the end plate of one nerve cell and the dendrite of the next.
1. An impulse arrives at the synapse
2. At the end plate are tiny sacs containing a chemical (**neurotransmitter**)
3. The chemical is released into the synaptic gap
4. The chemical diffuses across the gap and the impulse 'restarts' on the other side.

The reflex arc is the basic working unit of the nervous system and rapidly and automatically links stimuli to responses via the **integrators** of the central nervous system.

SPINAL CORD

Grey matter

White matter

Relay neurone: carries impulse across spinal cord. This allows the action to be modified by impulses carried down the spinal cord from the brain.

Receptor: a cell or organ which receives a stimulus and converts it into impulse.

Sensory neurone: carries impulse from the receptor to the central nervous system.

Spinal nerve: has both sensory and motor neurones, and leaves spinal column between adjacent vertebrae.

Motor neurone: carries impulse from the central nervous system to the effector.

Effector: muscle or gland which carries out an action (response) to 'deal with' the initial stimulus.

SENSE ORGANS ARE GROUPS OF RECEPTOR CELLS

ORGAN	RESPONDS TO	'SENSE'
Eye	light	sight
Ear	sound	hearing
Nose/tongue	chemicals	smell/taste
Skin	touch	touch/feeling
Skin/part of brain	temperature	hot and cold

Muscle-bone machines
are responsible for movement.

Antagonistic pairs of muscles are necessary for controlled movement at a joint. Muscles may only exert a force by **contraction**. To reverse a muscular movement therefore requires contraction of an opposing (antagonistic) muscle.

Bending the arm:
Biceps contracts
Triceps relaxes

Straightening the arm:
Triceps contracts
Biceps relaxes

So, the biceps is a **flexor** and the triceps is the **extensor** of the elbow joint, and biceps and triceps make up an **antagonistic pair**.

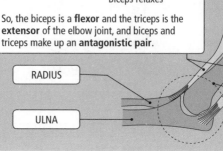

Tendons connect muscle to bone. They are **inelastic** so that muscle contraction can be converted into movement of bone.

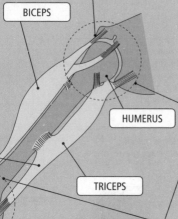

BICEPS

HUMERUS

RADIUS

ULNA

TRICEPS

CONTRACTION OF MUSCLE REQUIRES
- the **supply** of **glucose** and **oxygen** to release energy, in the form of ATP, by respiration
- the **removal** of **carbon dioxide** and **heat** (if respiration becomes **anaerobic** there must also be the removal of **lactic acid** or cramp may result)

These requirements are satisfied by the development of a system of capillary beds in the muscle together with their arteries and veins.

- a **stimulus** in the form of a **nervous impulse** delivered from the end plates of **motor neurones**.

Synovial joints are freely-moving because of the arrangement of the synovial membranes/cartilage between the ends of adjacent bones.

Homeostasis: controlling body temperature

Homeostasis is the process whereby the body adjusts to changes to keep essential internal conditions steady.

Many of the body's systems work best under specific chemical and physical conditions. To keep things in a steady state, the body uses a system of **detectors** and effectors. The detectors detect changes inside the body. The effectors then bring about changes in the opposite direction in order to restore equillibrium. Keeping the body's internal temperature at 37°C is a good example of **homeostasis**.

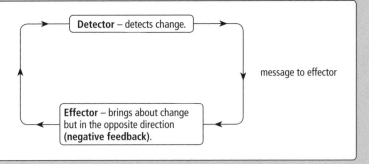

Detector – detects change.

message to effector

Effector – brings about change but in the opposite direction **(negative feedback)**.

BODY TOO HOT...

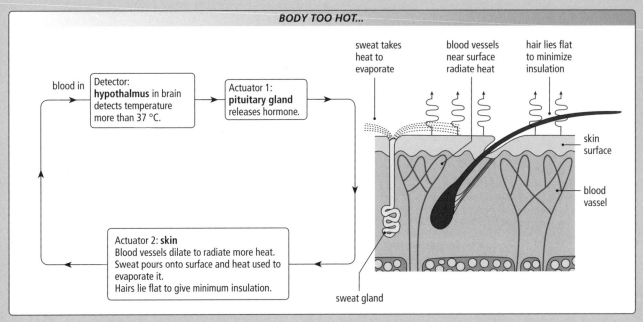

blood in

Detector: **hypothalmus** in brain detects temperature more than 37 °C.

Actuator 1: **pituitary gland** releases hormone.

Actuator 2: **skin**
Blood vessels dilate to radiate more heat.
Sweat pours onto surface and heat used to evaporate it.
Hairs lie flat to give minimum insulation.

sweat takes heat to evaporate

blood vessels near surface radiate heat

hair lies flat to minimize insulation

skin surface

blood vassel

sweat gland

BODY TOO COLD...

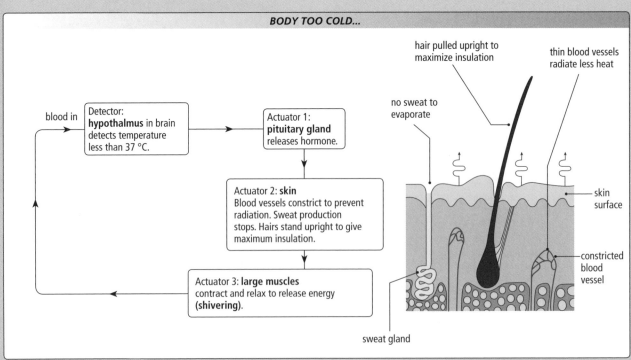

blood in

Detector: **hypothalmus** in brain detects temperature less than 37 °C.

Actuator 1: **pituitary gland** releases hormone.

Actuator 2: **skin**
Blood vessels constrict to prevent radiation. Sweat production stops. Hairs stand upright to give maximum insulation.

Actuator 3: **large muscles**
contract and relax to release energy **(shivering)**.

hair pulled upright to maximize insulation

thin blood vessels radiate less heat

no sweat to evaporate

skin surface

constricted blood vessel

sweat gland

Drug abuse

A drug is any substance taken into the body that modifies or affects chemical reactions in the body.

ANTIBIOTICS ARE USEFUL DRUGS!
These substances are produced by *molds* and either
- kill bacteria, or
- slow down reproduction of bacteria so that the body's own immune system destroy them. *Penicillin* is a well-known example of an antibiotic.

HEROIN
Causes lethargy, loss of self-control, mental and physical deterioration.

There is a serious danger of overdose, leading to coma and death.

Addiction can lead to crime as a way to obtain the drug ('extra' for depressions).

Sharing needles for injection of heroin can lead to infection eg HIV/AIDS and hepatitis.

TOBACCO
Nicotine is addictive. Tar and smoke particles from cigarette smoke contain chemicals' which cause cancer carbon monoxide damages the cilia in the lungs, leading to infection and breathing difficulties including **bronchitis** and **emphysema**.

ALCOHOL
Alcohol is a depressant. It affects judgement and self-control. This can cause accidents, especially when driving.

Excessive alcohol intake can cause **hepatitis** (liver inflammation) or **cirrhosis** (permanent liver scarring).

It can lead to higher blood pressure and make **coronary heart disease** more likely.

PAIN KILLERS
Aspirin and paracetamol are **useful** drugs but when taken is excess they are dangerous.

Paracetamol can lead to liver failure and death.

Aspirin can cause bleeding of the stomach.

1. The diagram shows a single neuron.

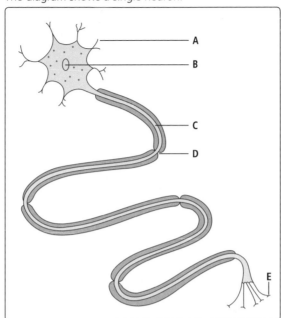

Which of the parts labelled A to F:
a. connects with another neuron? (1)
b. insulates the axon to limit interference with other neurones? (1)
c. contains high concentrations of chemicals called neurotransmitters? (1)
d. connects with an effector, such as a muscle? (1)
e. allows a nerve impulse to 'jump' quickly along the axon? (1)
f. contains DNA? (1)

2. The bar graph shows the units of alcohol present in some different types of alcoholic drink. The line graph relates alcohol level in the blood to the units of alcohol consumed.

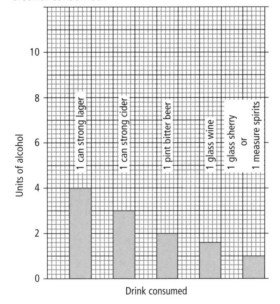

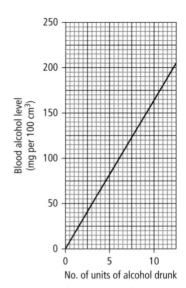

a. How many units of alcohol are present in one can of strong lager? (1)
b. How much alcohol would be present in the blood after drinking a pint of bitter and two measures of whisky? (2)
c. The UK limit for blood alcohol level in a person driving a motor vehicle is 80 mg per 100 cm^3. A person drank three cans of cider and one glass of wine – by how much would her blood alcohol level be over the legal limit? Show your working. (3)
d. Why is it dangerous to drive when over the 'legal limit'? (2)
e. Alcohol is cleared from the bloodstream by the action of the liver, at the rate of about 1 unit per hour. How many hours after drinking the alcohol described in part (**c**) would a person 'safe' to drive? (2)
f. Name the disease of the liver caused by excessive consumption of alcohol. (1)

3. Each year in Britain as many as 750 people are paralysed as a result of damage to the spinal cord.

a. The pie chart shows the different causes of this damage.
    The pie chart shows the causes of damage to the spinal cord.

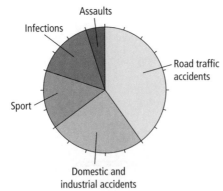

i.   What is the most likely cause of damage?   (1)
ii. What proportion of injuries to the spinal cord are caused by domestic and industrial accidents? Show your working.   (2)
iii. Why is it important that young people are vaccinated against meningitis?   (1)

b. Explain how it is possible that a girl with a damaged spinal cord cannot feel the shoes she is wearing but is able to write quite easily.   (2)

4. The diagram shows a section of the spinal cord, illustrating the pathway of a simple reflex arc.

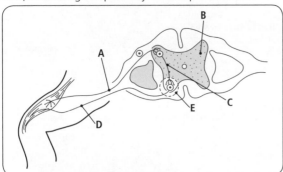

a. Identify the structures labelled A to E.   (5)
b. Draw a simple diagram to explain how impulses cross the gap labelled F.   (3)
c. Some drugs have their effects by affecting natural chemicals found in this gap. Several of these drugs are described in the table below.

	DRUG	EFFECT ON GAP
A	Caffeine	Prevents breakdown of natural chemical after it has crossed gap
B	Marijuana	Blocks natural chemical before it can cross gap
C	Heroin	Mimics (copies) natural chemical
D	Cocaine	Prevents removal of natural chemical after it has crossed the gap
E	Nicotine	Mimics (copies) natural chemical

State whether each of these drugs will act as a stimulant or as a sedative, Explain your answer.   (5 × 2)

5. A group of students suggested that coffee is enjoyable because it speeds up the heart rate. They gave several groups of people different amounts of coffee one morning and collected the following information.

NUMBER OF CUPS OF COFFEE DRUNK	HEART RATE (BEATS PER MINUTE) TOTAL	MEAN
0	74, 76, 72, 72, 78, 68	
1	78, 78, 82, 72, 72, 70	
2	78, 78, 79, 87, 80, 72	
3	80, 82, 78, 81, 78, 76	
4	76, 78, 88, 90, 88, 86, 78	
5	80, 90, 88, 88, 94, 92	

a. i.   Copy and complete the table by calculating the mean heart rate for each of the test groups.
    ii. Present your data in a suitable graphical form.
    iii. Does this data support the hypothesis that coffee affects the heart rate?
    iv. Suggest three precautions that the students should have taken to ensure that their data were valid.

b. How could you use epidemiology to investigate the hypothesis that heroin dependence in women leads to lower birth mass of children? Why is an epidemiological approach necessary to use in this sort of investigation?

6. Study the diagram below, then answer the questions that follow.

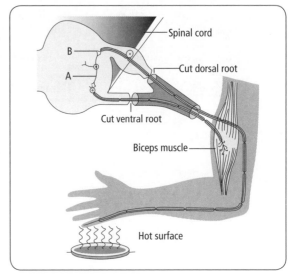

a. Name the type of neurone labelled **A**. What important function does it have?
b. Name the gap labelled **B**.

c. In what form is a message transmitted:
   i. along a nerve fibre
   ii. across the gap **B**?
d. Give two examples of spinal reflexes, two of cranial reflexes and one conditioned reflex.
e. How would sensation in the limb be affected if:
   i. the dorsal root (branch) was cut
   ii. the ventral root was cut?
f. Would a reflex action occur in the limb when the dorsal root was being cut? Explain your answer.
g. Name the parts C to G on the diagram of a sensory neurone below. State two ways in which this neurone differs from a motor neurone.

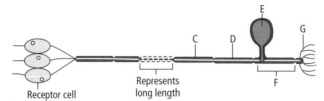

Receptor cell

Represents long length

REVISION SUMMARY:

Match the words with the correct definitions

	TERM		DEFINITION
A	Neurone	1	Junction between nerve cells
B	Stimulus	2	An electrical insulator for the neurone
C	Synapse	3	A structure which can detect a stimulus
D	Neurotransmitter	4	An automatic action with a positive survival value
E	Myelin	5	A conducting cell
F	Receptor	6	A structure which can carry out an action
G	Effector	7	Region of brain which matches stimuli to responses
H	Reflex	8	The brain and spinal cord
I	Association centre	9	A change in the environment
J	CNS	10	Chemical which bridges the synapse

The eye is a sense organ

Different tissues work together to perform one function.

- **Retina** contains light sensitive cells.
- **Cornea** and **lens** focus the light onto the retina.
- **Sclera, choroid** and **iris** protect and support the eye.

Accommodation: is the process of producing a finely focused image on the retina. It is carried out by the action of the **ciliary muscles** on the **lens**.

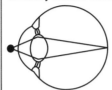 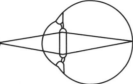

Close object:
Light must be **greatly refracted** (bent) ciliary muscles **contract**, pull eyeball inwards ligaments **slack** lens becomes **short and fat**.

Distant object:
light can be **less refracted** ciliary muscles **relax**, eyeball becomes spherical ligaments **tight** lens is pulled **long and thin**.

Rods: provide **black-and-white** images but, because several may be 'wired' to a single sensory neurone in the optic nerve, they provide **great sensitivity** at low light intensity (night vision), but images lack detail.

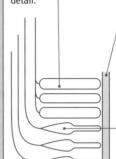

Layer of pigment: prevents internal reflection which might lead to 'multiple/blurred' images.

Cones: provide very **detailed** images, in **colour** (there are three types, sensitive to red, green and blue light) **but only under high light intensity** (their connections to the optic nerve make them rather insensitive).

| Lens | Ciliary muscle | Suspensory ligament |

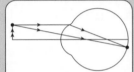

Cornea: a transparent layer that is responsible for most of the refraction (bending) ol light rays from the object to form the inverted, and smaller image on the retina.

Pupil: the circular opening that lets light into the eye. Appears 'black' because the choroid is visible through it.

Iris: the 'coloured' part of the eye, which may expand and contract to control the amount of light which enters the eye – this is a **reflex action**.

Low light intensity: radial muscles ol iris contract and the pupil is opened wider – **more light may enter and reach retina.**

High light intensity: circular muscles of iris contract and the pupil is reduced in size – **less light may enter** and retina is protected from **bleaching.**

Sclera: the tough outer coat, which protects the eye against damage and provides attachment for the muscles which move the eye in its socket.

Choroid: a darkly coloured layer that (a) limits internal reflections and (b) contains blood vessels which help to nourish the cells of the retina.

Retina: contains the light-sensitive cells, the rods and cones. Light arriving at this layer will produce an **inverted, smaller image**.

Yellow spot (fovea): has the greatest density of cones and thus offers **maximum sharpness** but only works at full efficiency in **bright light**.

Optic nerve: composed of sensory neurones that carry nervous impulses to the **visual centre** at the rear of the brain.

Blind spot: corresponds to the exit point for the optic nerve. There are no light-sensitive cells here so that light falling on this region cannot be detected.

Aqueous humour: watery fluid that supports the cornea and the front chamber of the eye.

Vitreous humour: a jelly-like substance that helps to keep the shape of the eyeball, supports the lens and keeps the retina in place at the rear of the eye.

1. Jasmine went into a dark room from a bright corridor.

a. Jasmine's right eye before and after entering the dark room is shown

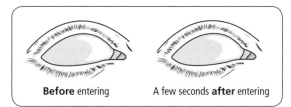

Before entering   A few seconds **after** entering

i. Copy and complete the diagrams by **drawing** the appearance of the pupil and iris
1 before entering the dark room, (1)
2 a few seconds after entering the dark room. (1)
ii. Label the following parts of the eye on the first diagram.
iris   pupil   sclera (3)

b. Explain how the size of the pupil was changed when Jasmine went into the dark room. (2)

c. Explain why Jasmine could see shapes but **not** colours in the dark room. (3)

CIE 0610 November '05 Paper 3 Q4

2. The diagram shows a section through the human eye.

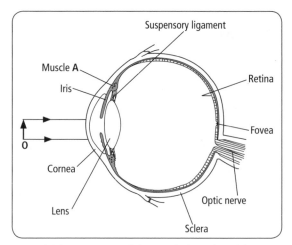

a. Which **two** parts of the eye help to bend the light rays to bring them to a focus? (1)

b. If object **O** were moved closer to the eye, what would muscle **A** do and how would this help to bring the light to a focus? (2)

3. Match the terms with their definitions.

	TERM		DEFINITION
A	Retina	1	Receptors that can distinguish red from green from blue
B	Rods	2	Structure that adjusts, focusing light onto the retina
C	Cones	3	Muscular structure that adjusts the amount of light entering the eye
D	Cornea	4	The opening that allows light to reach the retina
E	Lens	5	The light-sensitive layer in the eye
F	Optic nerve	6	Muscular structure that allows accommodation
G	Pupil	7	Exit point for the optic nerve
H	Iris	8	Carries sensory information to the visual association centre
I	Ciliary body	9	Receptors that are sensitive to low levels of light
J	Blind spot	10	Transparent layer that refracts light

(10)

4. Below is a section through the eye with a ray of light passing through it and four muscles labelled **A**, **B**, **C** and **D**.

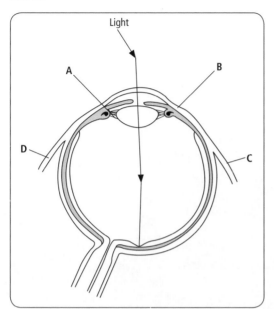

Light

A

B

D

C

a. Complete the table.

PART	NAME OF MUSCLE	EFFECT OF CONTRACTION
A		allows the lens to become fatter for focusing on close objects
B	iris circular muscle	

(2)

Muscles **C** and **D** are voluntary muscles that are antagonistic. They are attached to the eye socket of the skull.

b. i. Explain the terms *voluntary* and *antagonistic*.

voluntary

antagonistic (2)

ii. Suggest the effect on the eye when muscle **C** contracts. (1)

iii. Explain how the eye would return to its original position after this contraction. (2)

c. Light passes through parts of the eye to reach the retina.

Complete the flow chart by putting the following terms in the boxes to show the correct order that the light passes through them.

aqueous humour   cornea   lens   pupil   vitreous humour

☐ → ☐ → ☐ → ☐ → ☐ → retina

(2)

d. The retina contains rods and cones.

Complete the table to distinguish between rods and cones.

	TYPE OF LIGHT DETECTED	DISTRIBUTION IN THE RETINA
Rods		
Cones		

(4)

CIE 0610 June '05 Paper 3 Q2

5. Below is a section through the retina of a human eye.

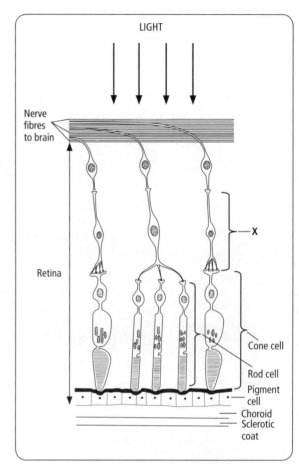

LIGHT

Nerve fibres to brain

Retina

X

Cone cell

Rod cell

Pigment cell

Choroid

Sclerotic coat

a. i. Describe the exact function of light receptors in terms of conversions they bring about. (2)

ii. Describe how the functions of the **rods** are different from those of the **cones**. (2)

b. What **type** of cell is **X** and how is its structure related to its function? (4)

c. Through what structures will light rays pass from the outside of the eye until they stimulate the rods/cones. (3)

d. A bright light is suddenly shone into a person's eyes. The person raises his fore-arm in order that his hand will shield his eyes from the glare.

i. Describe how antagonistic muscles in the arm will bring about this response. (5)

ii. Describe how the eye is linked with the effector to produce this response. (5)

RECEPTORS AND THE SENSES Crossword

ACROSS:
2   Light-sensitive layer of the eye
5   Sense that depends on the semi-circular canals
6   The organ responsible for 9 across
7   A reflex action that might be set off by touch receptors in the trachea
9   The sense of vision
11 Photoreceptors responsible for colour vision
12 Tighten during viewing of a distant object
15 Automatic action with a positive survival value
18 Photoreceptors that are most common around the edge of the retina
19 Location of chemoreceptors responsible for the sense of smell
22 A sense with receptors located in the skin
25 With 1 down – a structure responsible for detecting changes in temperature
26 Receptor that starts the knee jerk reflex
28 Effector responsible for controlling the shape of the lens (7, 6)
31 A change in the environment that can be detected by a receptor
32 Location of receptors concerned with touch and heat
33 Muscle that controls the amount of light entering the eye
34 Does this layer give you a black eye?

DOWN:
1   Structure capable of detecting a stimulus
3   Location of the chemoreceptors responsible for taste
4   Crystalline structure that 'fine focuses' light onto the retina
8   Is this spot yellow?
10 Sense related to the ears
13 The ability to link a sensory input with the correct motor output
14 There are four separate types of this sense, all strated by receptors on the tingue
16 A chemical sense that helps to protect the breathing passages from dangerous chemicals
17 Sensory nerve from the ear
20 Sensory nerve from the eye
21 Has outer, middle and inner sections – all concerned with responding to sound stimuli
23 Transparent, refractive layer at the front of the eye
24 No photoreceptors here!
26 This type of neurone leaves a sense organ
27 Protective secretion for the eyes – especially when peeling onions
29 Sensation caused by overstimulation of any receptor
30 Circular opening to admit light to the rear of the eye

Hormones and plant growth

Groups of plant hormones called **auxins** control the growth of shoots and roots.

PHOTOTROPISM

Plant shoots grow towards light. This is called phototropism. Phototropism is caused by auxins produced in the shoot tip. The auxins promote growth and are redistributed under the influence of light.

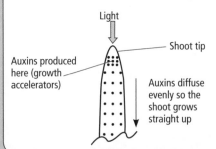

Light

Shoot tip

Auxins produced here (growth accelerators)

Auxins diffuse evenly so the shoot grows straight up

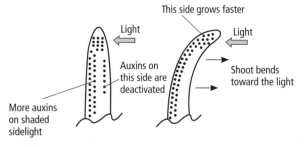

This side grows faster

Light

Light

Auxins on this side are deactivated

Shoot bends toward the light

More auxins on shaded sidelight

GEOTROPISM

Plant roots grow downwards in the direction of the pull of gravity. This is called geotropism. Geotropism is caused by auxins produced in the root tip. Root auxins slow down growth (i.e. opposite to the effect of shoot auxins).

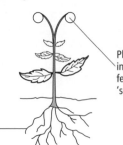

Auxins evenly spread so the root grows straight down

Auxins produced here

Root cap

Pull of gravity

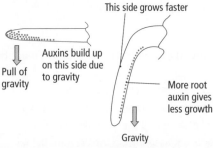

This side grows faster

Auxins build up on this side due to gravity

Pull of gravity

More root auxin gives less growth

Gravity

USING PLANT HORMONES

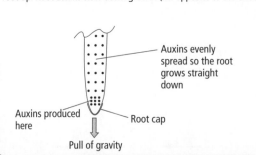

Hormone rooting powders can promote root growth in plant cuttings.

Plant hormones can 'trick' flowers into forming fruits without fertilization? this produces 'seedless' fruits.

Selective weed killers are absorbed by broad-leaved weeds but narrow-leaved grasses and cereal crops absorb less. Hormones cause the weeds to grow so quickly that they wither and die, and the crop of cereals can grow better without competition.

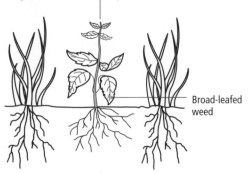

Broad-leafed weed

1. All living organisms respond to changes in their environment. The responses they show have an important effect on their survival. Plants respond by altering their direction of growth. Because growth is involved, plants respond more slowly than animals do.

Complete the following sentences about plant responses.

a. A tropism is defined as a change in the direction of .... of a plant in response to a directional ....

b. In geotropism, the plant is responding to ....

c. The shoot of a plant will grow towards light, that is it shows .... phototropism. This response allows the plant to produce more food by .... (5)

2. Growing oat seedlings are placed in a box that only allows light from one side to reach them, as shown in the diagram below.

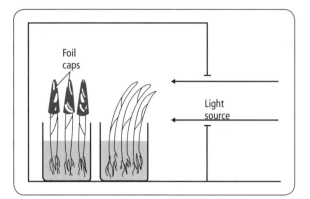

Which of the following conclusions is the correct one? Explain your answer. (2)

The shoot with the tinfoil cap:

a. cannot photosynthesize without light, and so cannot grow in a curved shape
b. is too far away from the light to be stimulated
c. cannot respond because the tips must receive the stimulus
d. is deprived of carbon dioxide, and so cannot grow

3. Young oat seedlings were treated with a substance X. The substance X was given to the seedling by mixing it with lanolin paste and smearing the paste on one side of the seedling, as shown in the diagram below.

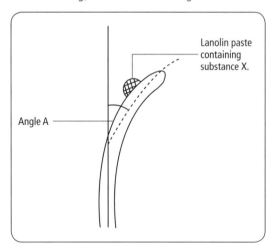

An experiment was carried out in which different concentrations of substance X were given to different seedlings, and the angle A measured in each case. The results are shown below.

CONCENTRATION OF X / mg dm⁻³	ANGLE A / DEGREES
1	3
2	6
4	10
7	16
10	19

a. Plot these results in the form of a graph. (5)
b. What would be the value of angle A at a concentration of X of 5 mg dm⁻³? (2)
c. What is the name given to the response shown by the oat seedlings? (1)
d. What is the name of substance X? (1)

4. Auxin is a hormone made by the tips of plant shoots.
The diagram below shows the movement of auxin in two young shoots, **A** and **B**, which were treated in different ways. '**X**' shows where auxin was made.
Both shoots were kept in the dark.

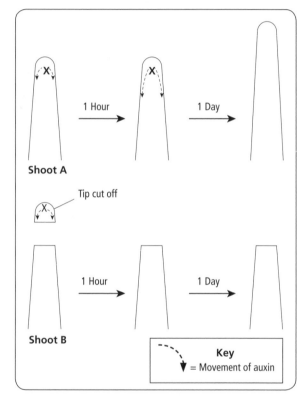

a. Explain the difference in the growth of shoot **A** and shoot **B** at the end of one day. (4)
b. **A** third shoot, **C**, was grown in a box so that light shone onto it from only one side. The diagram below shows movement of auxin in this shoot and the result of the experiment. (1)

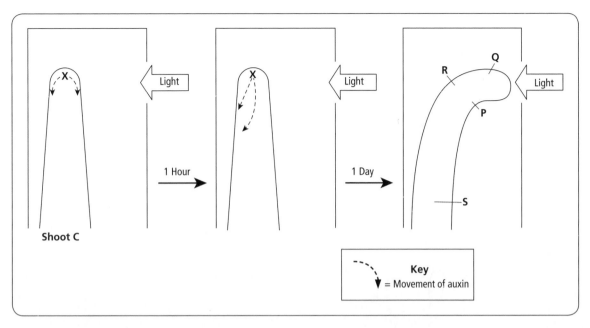

Shoot C

1 Hour →

1 Day →

Key
↓ = Movement of auxin

i.   Describe the movement of auxin in shoot **C** after one hour.

ii. Auxin causes plant cells to elongate (grow longer). At which point, **P**, **Q**, **R** or **S**, would cells have elongated the most?
     Draw a ring around one answer.

        P   Q   R   S   (1)

c. Plant hormones are sometimes used by humans to control plant growth. Give **two** examples of this.   (2)

5. a. State **three** conditions necessary for the germination of most seeds.   (3)

A student carried out an experiment on the direction of growth of the root of a germinating seed and the shoot of a seedling. Figure **A** shows the experiment when first set up. The electric motors slowly turn the cork base and the plant pot Figure **B** shows the experiment after two days. The root and shoot received the same amount of light from all directions.

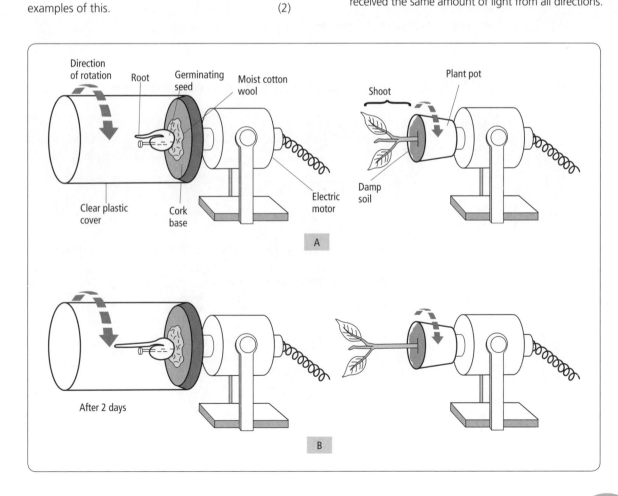

A

B

b. Suggest why a clear plastic cover was provided for
    the root but not for the shoot.   (1)
c. i. Complete the diagram below to show how the
        root and shoot would appear 24 hours **after
        the motors were switched off**.

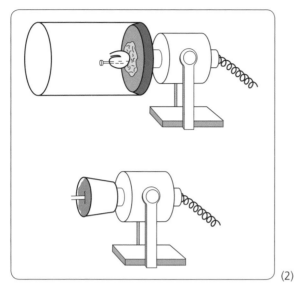

                                                        (2)
    ii. Name the responses shown by the root and the
        shoot that you have drawn and explain how the
        responses have come about.
        Root response
        Explanation
        Shoot response
        Explanation   (4)
    iii. Suggest why these responses were **not**
         shown in Figure A.   (1)

REVISION SUMMARY: Match the terms with the correct definition

	TERM		DEFINITION
A	Tropism	1	Form of chemical pest control, allowing crop plants to grow without competition
B	Auxin	2	Fusion of male and female gametes
C	Gravity	3	Plant response to light
D	Geotropism	4	Hormone that increases the rate of root growth from the base of a cut stem
E	Phototropism	5	Directional growth response in a plant
F	Apex	6	Downward force resulting from the mass of the Earth
G	Rooting powder	7	Hormone affecting growth of both root and shoot
H	Weedkiller	8	Development of a seed into a young plant
I	Germination	9	Plant response to gravity
J	Fertilization	10	The tip of a root or shoot

Chapter 26:

Human reproduction and growth

MALE REPRODUCTIVE SYSTEM

- produces **testosterone**
- manufactures **male gametes**
- transfers male gametes to female

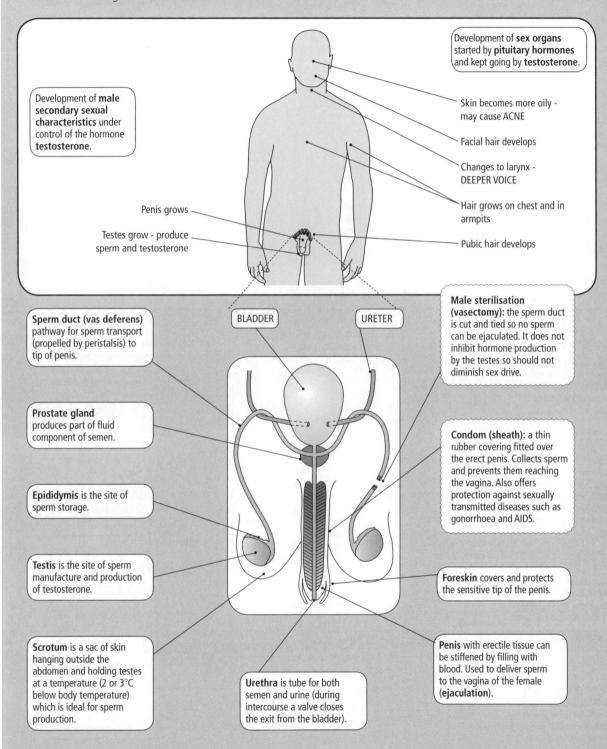

Development of **sex organs** started by **pituitary hormones** and kept going by **testosterone**.

Skin becomes more oily - may cause ACNE

Facial hair develops

Changes to larynx - DEEPER VOICE

Hair grows on chest and in armpits

Pubic hair develops

Development of **male secondary sexual characteristics** under control of the hormone **testosterone**.

Penis grows

Testes grow - produce sperm and testosterone

BLADDER

URETER

Sperm duct (vas deferens) pathway for sperm transport (propelled by peristalsis) to tip of penis.

Prostate gland produces part of fluid component of semen.

Epididymis is the site of sperm storage.

Testis is the site of sperm manufacture and production of testosterone.

Scrotum is a sac of skin hanging outside the abdomen and holding testes at a temperature (2 or 3°C below body temperature) which is ideal for sperm production.

Male sterilisation (vasectomy): the sperm duct is cut and tied so no sperm can be ejaculated. It does not inhibit hormone production by the testes so should not diminish sex drive.

Condom (sheath): a thin rubber covering fitted over the erect penis. Collects sperm and prevents them reaching the vagina. Also offers protection against sexually transmitted diseases such as gonorrhoea and AIDS.

Foreskin covers and protects the sensitive tip of the penis.

Penis with erectile tissue can be stiffened by filling with blood. Used to deliver sperm to the vagina of the female **(ejaculation)**.

Urethra is tube for both semen and urine (during intercourse a valve closes the exit from the bladder).

FEMALE REPRODUCTIVE SYSTEM

- produces **oestrogen / progesterone**
- develops **female gametes**
- accepts sperm
- allows development / birth of foetus

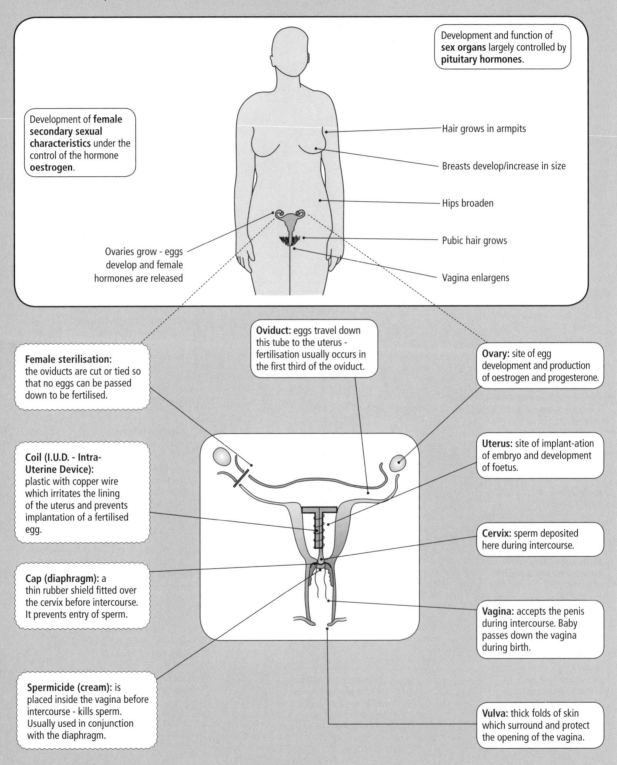

Development and function of **sex organs** largely controlled by **pituitary hormones**.

Development of **female secondary sexual characteristics** under the control of the hormone **oestrogen**.

Hair grows in armpits

Breasts develop/increase in size

Hips broaden

Pubic hair grows

Vagina enlargens

Ovaries grow - eggs develop and female hormones are released

Oviduct: eggs travel down this tube to the uterus - fertilisation usually occurs in the first third of the oviduct.

Ovary: site of egg development and production of oestrogen and progesterone.

Female sterilisation: the oviducts are cut or tied so that no eggs can be passed down to be fertilised.

Coil (I.U.D. - Intra-Uterine Device): plastic with copper wire which irritates the lining of the uterus and prevents implantation of a fertilised egg.

Uterus: site of implant-ation of embryo and development of foetus.

Cervix: sperm deposited here during intercourse.

Cap (diaphragm): a thin rubber shield fitted over the cervix before intercourse. It prevents entry of sperm.

Vagina: accepts the penis during intercourse. Baby passes down the vagina during birth.

Spermicide (cream): is placed inside the vagina before intercourse - kills sperm. Usually used in conjunction with the diaphragm.

Vulva: thick folds of skin which surround and protect the opening of the vagina.

The menstrual cycle

is a regular series of changes to the female reproductive system in preparation for fertilization and pregnancy.
It is controlled by **hormones** from the **pituitary gland** and the **ovary**.

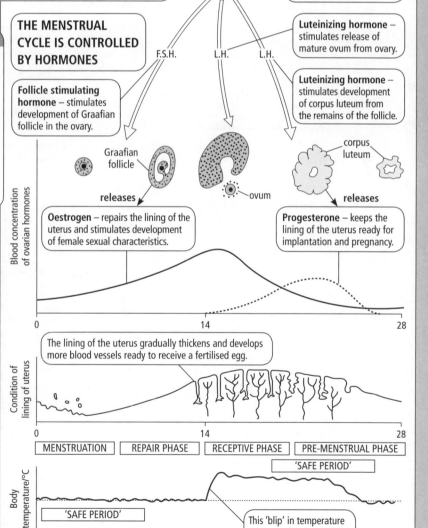

In humans the cycles of the two ovaries are out of phase, so **each ovary** ovulates every 56 days but **each female** ovulates (releases an egg) every 28 days.

The uterus lining begins to degenerate **unless embryo implantation** has occurred when **progesterone** (from the corpus luteum) keeps the lining intact to begin pregnancy.

Uterus lining and its blood vessels are well-developed to receive an embryo. This optimum set of conditions remains for 6–7 days after ovulation.

A TYPICAL MENSTRUAL CYCLE

Stage 1 MENSTRUATION

Stage 2 REPAIR PHASE (of uterus lining)

Stage 3 RECEPTIVE PHASE

Stage 4 PRE-MENSTRUAL PHASE

Separation of uterus lining – blood and fragments of tissue leave the body via the vagina. This monthly 'period' occurs only in primates such as humans.

Ovulation, the release of the egg from the Graafian follicle into the oviduct. This is stimulated by **luteinizing hormone** and the release of fluid and an associated 'blip' in body temperature (see right) means that some human females are aware that ovulation has occurred.

Brain – receives information from other parts of the body, processes it and then 'instructs' pituitary gland.

Pituitary gland – releases hormones which control activity of the ovary.

Luteinizing hormone – stimulates release of mature ovum from ovary.

Luteinizing hormone – stimulates development of corpus luteum from the remains of the follicle.

CONTRACEPTION AND THE MENSTRUAL CYCLE

- **The pill:** an oral dose of one or both hormones **oestrogen** and **progesterone** that acts as a **feedback inhibitor** of the release of **luteinizing** hormone by the pituitary gland. Thus **ovulation cannot occur** and **no pregnancy can result**.

- **The rhythm method** assumes that the time up to two days before ovulation and from four days after ovulation are **safe**, i.e. there should be no eggs available to be fertilized. This method is unreliable.

THE MENSTRUAL CYCLE IS CONTROLLED BY HORMONES

F.S.H.   L.H.   L.H.

Follicle stimulating hormone – stimulates development of Graafian follicle in the ovary.

Graafian follicle

releases

ovum

corpus luteum

releases

Oestrogen – repairs the lining of the uterus and stimulates development of female sexual characteristics.

Progesterone – keeps the lining of the uterus ready for implantation and pregnancy.

Blood concentration of ovarian hormones

The lining of the uterus gradually thickens and develops more blood vessels ready to receive a fertilised egg.

Condition of lining of uterus

| MENSTRUATION | REPAIR PHASE | RECEPTIVE PHASE | PRE-MENSTRUAL PHASE |

'SAFE PERIOD'

Body temperature/°C

'SAFE PERIOD'

This 'blip' in temperature corresponds to ovulation.

The placenta

is responsible for protection and nourishment of the developing fetus.

PLACENTA: this disc-shaped organ has a number of functions:
- exchange of soluble materials between mother and fetus
- physical attachment of the fetus to the wall of the uterus
- protection
  1 of fetus from mother's immune system
  2 against dangerous fluctuations in mother's blood pressure
- hormone secretion. These hormones keep the wall of the uterus in 'pregnancy' state as the corpus luteum breaks down by the 3rd month.

The placenta is lost as the 'afterbirth' following birth of the fetus.

Wall of uterus: this is very muscular. At full term of the pregnancy a hormone (**oxytocin**) is secreted from the pituitary gland and makes this muscle contract in a series of waves to expel the fetus. The same hormone is used in the intravenous drip that includes birth when the pregnancy has gone on for too long.

Fetus: this develops from a **zygote**.

CELL DIVISION

ZYGOTE at fertilization

BALL OF CELLS at implantation

CELL DIVISION AND MOVEMENT

EMBRYO

CELL DIVISION, MOVEMENT AND SPECIALIZATION

Recognisably human form by end of eighth week: FETUS

Amnion: the membrane that encloses the amniotic fluid. This is 'ruptured' just before birth.

Amniotic fluid: protects the fetus against
- mechanical shock
- drying out
- temperature fluctuations

Some of the fetal cells fall off into this fluid and can be collected by **amniocentesis**. The cells can be analysed to detect disease, genetic abnormalities and even the sex of the fetus.

UMBILICAL CORD: contains blood vessels which carry materials which will be/have been exchanged between mother and fetus. The cord connects the fetus to the placenta.

Mucus plug in cervix: protects foetus against possible infection. The plug is expelled just before birth.

Placental villi – increase the surface area for exchange.

PASSING FROM MOTHER TO FETUS

Soluble nutrients, e.g. glucose and amino acids

Oxygen

Antibodies

BUT also

Viruses, e.g. HIV

Nicotine/heroin

and FROM FETUS TO MOTHER

Carbon dioxide

Urea

'Pit' in wall of uterus contains mother's blood. N.B. there is **no direct contact** between maternal and foetal blood.

Umbilical artery from fetus to placenta – deoxygenated blood containing waste.

Umbilical vein from placenta to fetus – oxygenated blood cleared of waste.

Counter-current flow system blood in maternal and foetal blood vessels **flows in opposite directions**. This gives the maximum possible area over which concentration gradients favour diffusion.

GROWTH AND DEVELOPMENT OF THE FETUS

MASS/g
3000
2000
1000

Internal organs all present

Birth usually occurs at about this time

Fetus fully formed, even fingerprints!

Fetus now has a good chance of survival if born

0 4 8 12 16 20 24 28 32 36 40
AGE OF FETUS/weeks

- Complete period from fertilization to birth = **gestation period**.
- Cell division converts single cell (zygote) to 30 million in a newborn baby.
- Most rapid growth from the twelfth week. As much as 1500 × gain in mass in 20 weeks.

1. The diagram below shows the reproductive system of a human male.

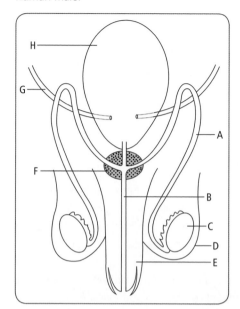

a. Name the parts labelled A, B, C, D and E.   (5)
b. Identify the part that produces the male gametes. (1)
c. Identify the part that produces a liquid part
    of semen.   (1)
d. Identify the part that produces testosterone.   (1)
e. Identify the part that also carries urine.   (1)
f. Identify the part that is cut during the procedure
    of vasectomy.   (1)

2. Arrange the following processes into the correct sequence necessary for the production of a human baby.
a. ejaculation
b. implantation
c. birth
d. fertilization
e. ovulation
f. development   (6)

3. The diagram below shows part of the human female reproductive system.

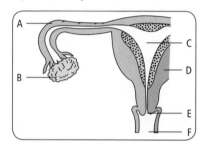

a. Identify the parts labelled A, B, C, D, E and F.   (6)
b. Make a simple copy of the diagram, and use an X to
    mark the place where fertilization occurs and a Y to
    show where implantation occurs.   (2)
c. Contraception prevents fertilization occurring. Use
    the diagram to explain how **(i)** the diaphragm and
    (ii) the pill act as contraceptives.   (2)

4. During pregnancy, the developing fetus is attached to the mother's body through the placenta. The diagram shows the structure of this organ, and its attachment to the fetus.

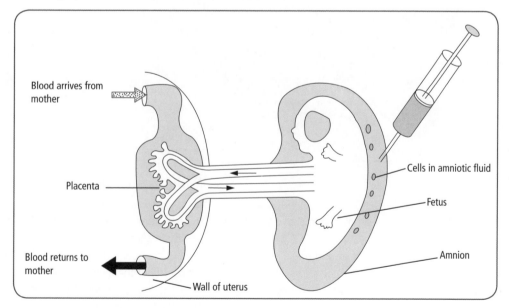

a. i.   Name **two** substances that pass from the fetus
         to the mother.   (2)
    ii. How is the structure of the placenta adapted to
         carry out this transfer?   (2)

    iii. Name **two** organs in the baby's body that take
         over the functions of the placenta once the
         baby is born. State the functions that these
         organs will perform.   (2)

b. A technique called amniocentesis can be used to check the genotype of the fetus. In this technique, a sample of amniotic fluid is removed using a long needle. The fluid contains cells shed from the body of the fetus. The cells can be stained, squashed and their chromosomes examined using a microscope.
    i.   What is the function of amniotic fluid? (1)
    ii. Which condition would the baby have if the microscopist counted an extra chromosome 21? (1)
c. Scientists have been able to count the number of these abnormal chromosomes in mothers of different ages. The results are shown in the table.

	AGE OF MOTHER/YEARS					
	20–24	25–29	30–34	35–39	40–44	45–49
Frequency of abnormal fetuses per 1000 females	0.5	2.8	4.9	9.9	19.8	27.2

i.   Plot this information in the form of a bar graph. (3)
ii. What is the percentage increase in abnormal foetuses between the age groups 30–34 and 45–49? Show your working. (2)
iii. Women may be given hormone replacement therapy to delay the onset of menopause. Give **one** possible benefit and **one** possible disadvantage of this treatment. (2)

5. In women two hormones control ovulation (the release of eggs from the ovaries). The drawing shows a monitoring machine which women can use each morning to measure the amounts of the two hormones. A test stick is dipped in the woman's urine, then placed in a slot in the machine.

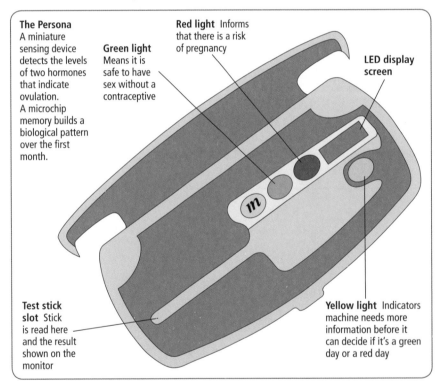

The Persona
A miniature sensing device detects the levels of two hormones that indicate ovulation. A microchip memory builds a biological pattern over the first month.

Green light Means it is safe to have sex without a contraceptive

Red light Informs that there is a risk of pregnancy

LED display screen

Test stick slot Stick is read here and the result shown on the monitor

Yellow light Indicators machine needs more information before it can decide if it's a green day or a red day

a. The machine monitors the levels of two hormones.
    i.   What is a hormone? (1)
    ii. How are hormones transported around the body? (1)
b. A woman is unlikely to become pregnant if she has sex on the days when the machine shows a green light during the test. Use information from the drawing to suggest why. (1)
c. The two hormones detected by the machine are oestrogen and LH.
    Explain how oestrogen and LH are involved in the control of the menstrual cycle. (4)
d. Hormones can be used to control human fertility. Describe the benefits and problems that might arise from using hormones in this way. (4)

6. The drawing shows the human fetus at six different stages of development.

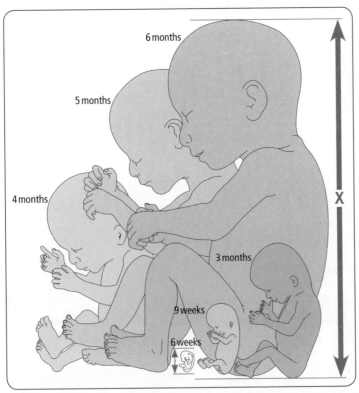

a. Apart from the increase in size, describe **one** change which occurs in the head and **one** change which occurs in the foot between the ages of **6 weeks** and **9 weeks**.
Use only features you can see in the drawing. (2)

b. i. Measure the length **X** on the drawings of the fetus at
   A 6-weeks
   B 6-months (2)

   ii. How many times larger is the 6-month fetus than the 6-week fetus? (1)

7. The diagram shows the reproductive organs of a woman.

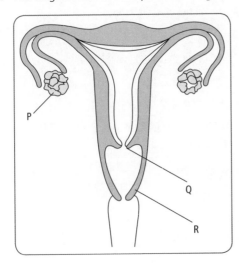

a. Label parts **P**, **Q** and **R**. (3)

b. What feature shown on the diagram might indicate that this woman is infertile? Explain your answer. (2)

This woman's infertility could be overcome by fertilizing one of her ova (eggs) in a test-tube.

c. i. Place an **X** on the diagram to show where the fertilised ovum should be implanted. (1)

   ii. Suggest **two** reasons why the fertilized ovum might be incubated for a few days, in a test-tube, before it is implanted. (2)

The diagram below shows the thickness of the uterus lining, and the level of progesterone in the blood, at different times in the woman's menstrual cycle. High levels of the hormone progesterone in the blood are needed to keep a thick spongy lining in the uterus.

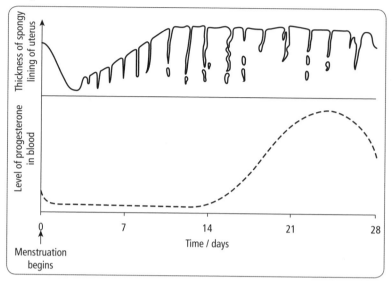

d. Using the information provided, suggest a time when implantation would be likely to be most successful. Explain your answer.

(2)

CIE 0610 June '97 Paper 3 Q5

REVISION SUMMARY: Matching pairs

Match each of the terms with the correct description or explanation.

	TERM		DESCRIPTION
A	Contraception	1	A developing mammal with recognizable internal and external organs
B	Embryo	2	Hormone that stimulates development of the corpus luteum
C	Fertilization	3	Hormone promoting the development of a ripe ovum
D	Fetus	4	Hormone that keeps the lining of the uterus ready for implantation and pregnancy
E	FSH	5	Tube for the transfer of ova from ovary to the site of fertilization
F	Gestation	6	The release of a female gamete
G	Implantation	7	The organ where materials are exchanged between maternal and fetal blood
H	LH	8	The fusion of male and female gametes to produce a zygote
I	Oestrogen	9	A developing human up to the eighth week of pregnancy
J	Ovary	10	Hormone that stimulates development of female secondary sexual characteristics
K	Oviduct	11	Hormone that stimulates production of sperm
L	Ovulation	12	Muscular chamber in which a young mammal develops
M	Placenta	13	Methods that prevent fertilization and pregnancy
N	Progesterone	14	Structure linking a fetus to the placenta
O	Testes	15	Attachment of the zygote to the lining of the uterus
P	Testosterone	16	The birth canal
Q	Umbilical cord	17	The site for the production of sperm and male sex hormone
R	Uterus	18	The period between conception and birth
S	Vagina	19	Tube that delivers sperm to the penis
T	Vas deferens	20	Site of production of ova and female sex hormones

Flower structure is adapted for pollination

CARPEL: the female part of the flower.

Stigma: surface on which pollen grains, containing male gametes, may be deposited.

Style: stalk that holds stigma in prominent position and down which pollen tube may grow.

Ovary: contains the ovule, which encloses the female gamete. Ovary wall may become part of the fruit.

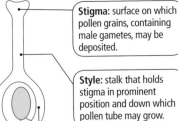

STAMEN: the male part of the flower.

Anther: produces pollen grains, containing male gametes, within the pollen sacs.

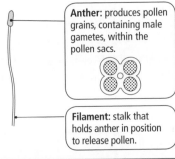

Filament: stalk that holds anther in position to release pollen.

PETALS: play an important part in pollination (see opposite).

SEPALS: protect the other floral parts against drying out and fungal attack. There are the same number as there are petals, and they are usually green in colour.

RECEPTACLE: the swollen tip of the flower stalk. It is the base on which the other parts of the flower stand.

POLLINATION is the transfer of pollen from the stamen to the stigma.

Cross-pollination involves pollen transfer between different flowers of the same species. It involves genetic variation but may be risky because it requires a **vector** (a carrier).

Self-pollination involves pollen transfer from stamen to stigma of the same flower. It is less risky but it limits the chances of genetic variation.

Wind-pollinated species usually occur in dense groups, e.g. the grasses.		Insect-pollinated species are usually solitary or in small groups.
Dull in colour. Small, or even absent to reduce obstruction of pollen access to stigma.	PETALS	Large, brightly coloured, may be scented and/or have guidelines. Base may produce attractive **nectar.**
Long and flexible so that pollen may be easily released.	STAMENS	Short and stiff to brush pollen against body of visiting insect.
Long and feathery, giving a large surface area to receive pollen.	STIGMA	Held inside petals to ensure contact with body of visiting insect.
Small, dry, enormous quantities.	POLLEN	Large, sticky, small amounts.

Bees are adapted to feed on pollen and nectar. They pollinate flowers when feeding.

Antennae detect scent of flower.

Mouthparts can suck nectar from flower.

Eyes detect colour of flower.

Wings permit movement between flowers.

Hairy abdomen may hold 'sticky' pollen.

Pollen brushes and baskets on third pair of legs collect pollen for larvae in hive.

1. The diagram below shows a bee that collects food materials from some flowers belonging to the same species. While it does this the bee also assists in the reproductive processes of the flowers.

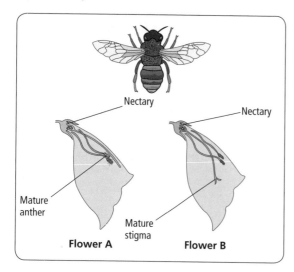

Flower A   **Flower B**

a. i. Name the stage in the reproduction of the plants in which the bee is involved. (1)

ii. Suggest how this process might take place between flowers **A** and **B**. (3)

b. The ovules in each flower can develop into seeds.

i. Which reproductive process must happen inside an ovule before it can become a seed? (1)

ii. State which part of the flower develops into a fruit. (1)

c. Explain why plants grown from the seeds produced by these flowers will be similar to each other but may not be identical. (4)

CIE 0610 November '05 Paper 2 Q6

2. A common garden weed, the Creeping Buttercup, can reproduce asexually by means of runners. The process is summarized in the diagram below.

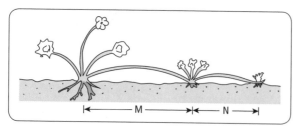

A horticultural student decided to investigate the growth of this weed. He took ten different buttercup plants and measured the lengths of the runners M and N. The results are shown in the table below.

PLANT NUMBER	LENGTH OF RUNNER M / mm	LENGTH OF RUNNER N / mm
1	175	140
2	210	130
3	305	130
4	320	170
5	300	125
6	170	120
7	230	135
8	300	125
9	260	145
10	200	100

a. Calculate the mean lengths for runners M and N. Show your working. (4)

b. Calculate the difference in mean length between runners M and N. (2)

c. Suggest why there are differences between the length M in the ten different plants. (2)

d. The student was trying to develop a weedkiller that could control the buttercups. He noticed that one of the plants was very resistant to the weedkiller. Explain how this method of reproduction would allow the buttercup to survive throughout the garden, even if weedkiller is used. (2)

3. a. What is the difference between asexual and sexual reproduction in terms of the number of parents involved and the amount of variation in the offspring? (2)

b. What are the advantages and disadvantages of these differences? (4)

c. Some flowering plants use runners to reproduce asexually. Describe how they do this. (4)

4. The figure below shows a whole flower of *Nicotiana* **(A)** and a section of the same flower **(B)**.

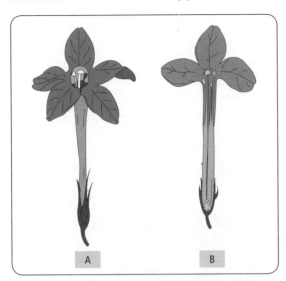

A   B

The figure below shows a whole flower of *Lilium* **(C)** and a section of the same flower **(D)**.

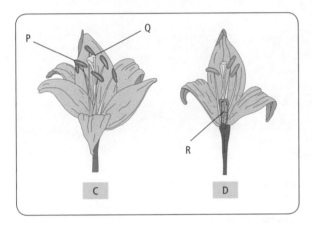

C   D

a. Complete the table below to show **five** differences between the *Nicotiana* flower and the *Lilium* flower. On each line you should compare **one** feature of the flowers. The first one is done for you.

	NICOTIANA FLOWER	LILIUM FLOWER
1	Fewer stamens are present	More stamens are present
2		
3		
4		
5		
6		

(5)

b. i. Compare the functions of structures labelled **P** and **Q**. (2)
   ii. Explain why reproduction in flowers is considered to be **sexual** rather than **asexual** reproduction. (2)

c. State **four** reasons why it is very unlikely that these flowers are wind-pollinated. (4)

REPRODUCTION IN PLANTS: Crossword

ACROSS:
4 Landing platform for male sex cells
7 What the ovary develops into after fertilization
9 An organism with both male and female sexual organs
10 Can be an agent of pollination or germination
12 Contains an embryo, a food store and a waterproof coat
13 Where the pollen is produced
15 Produces a nutritious, sugary secretion
17 Contains the female sex cells
18 The spreading out of offspring from a parent plant
19 The transfer of male sex cells to female flower parts
20 A flying pollinator
22 The main advantage of sexual reproduction
23 Stalk that supports 13 across in the male part of the flower
24 Underground stem for asexual reproduction – potato, for example

DOWN:
1 The female sex cell
2 Can be avoided if 18 across is efficient
3 A protective outer covering for a flower bud
5 Reproduction without fertilization
6 The fusion of male and female sex cells
8 Athletic structure for asexual reproduction?
11 May be coloured to attract insects
14 The devlopment of a seed into a young plant
16 Contains the male gamete
21 The pollen grain must send a tube down this structure

Chapter 28:
Germination and plant growth

Fertilization, fruits and seed germination

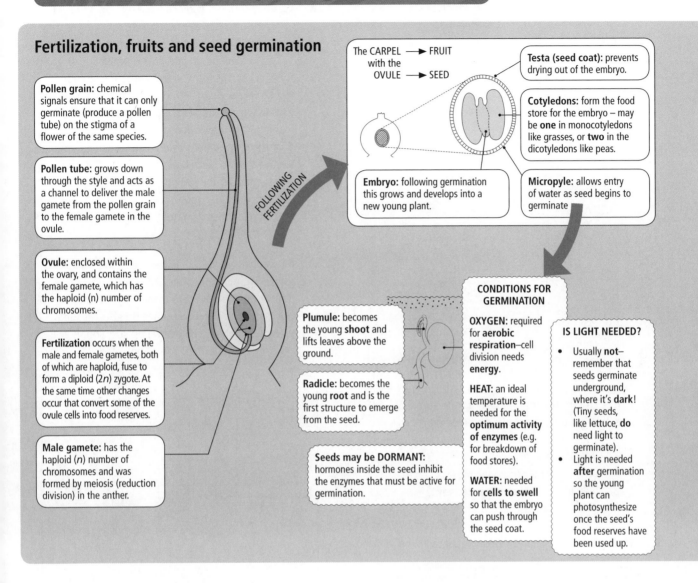

Pollen grain: chemical signals ensure that it can only germinate (produce a pollen tube) on the stigma of a flower of the same species.

Pollen tube: grows down through the style and acts as a channel to deliver the male gamete from the pollen grain to the female gamete in the ovule.

Ovule: enclosed within the ovary, and contains the female gamete, which has the haploid (n) number of chromosomes.

Fertilization occurs when the male and female gametes, both of which are haploid, fuse to form a diploid (2n) zygote. At the same time other changes occur that convert some of the ovule cells into food reserves.

Male gamete: has the haploid (n) number of chromosomes and was formed by meiosis (reduction division) in the anther.

FOLLOWING FERTILIZATION

The CARPEL → FRUIT
with the
OVULE → SEED

Testa (seed coat): prevents drying out of the embryo.

Cotyledons: form the food store for the embryo – may be **one** in monocotyledons like grasses, or **two** in the dicotyledons like peas.

Embryo: following germination this grows and develops into a new young plant.

Micropyle: allows entry of water as seed begins to germinate

Plumule: becomes the young **shoot** and lifts leaves above the ground.

Radicle: becomes the young **root** and is the first structure to emerge from the seed.

Seeds may be DORMANT: hormones inside the seed inhibit the enzymes that must be active for germination.

CONDITIONS FOR GERMINATION

OXYGEN: required for **aerobic respiration**–cell division needs **energy.**

HEAT: an ideal temperature is needed for the **optimum activity of enzymes** (e.g. for breakdown of food stores).

WATER: needed for **cells to swell** so that the embryo can push through the seed coat.

IS LIGHT NEEDED?

- Usually **not**– remember that seeds germinate underground, where it's **dark**! (Tiny seeds, like lettuce, **do** need light to germinate).
- Light is needed **after** germination so the young plant can photosynthesize once the seed's food reserves have been used up.

---

1. The diagram shows a section through half of a broad bean seed.

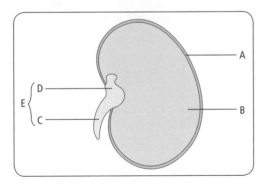

a. Name all of the parts labelled A to E on the diagram. (5)

b. Structure B contains a food store. A teacher explained that the food store provided energy for the growing seed. One student thought that the food would be fat, and another thought that it would be carbohydrate. Describe a test that you could carry out to decide who was right (or whether both were!). Name the reagents that you would use, the steps you would take and the results you might expect. (5)

2. Seeds will only germinate if the environmental conditions are suitable. The experiment shown below was set up to investigate the conditions required for germination.

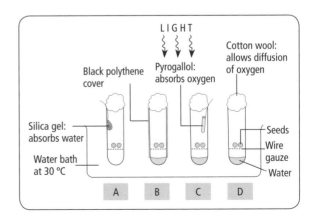

LIGHT

Black polythene cover

Pyrogallol: absorbs oxygen

Cotton wool: allows diffusion of oxygen

Silica gel: absorbs water

Water bath at 30 °C

Seeds
Wire gauze
Water

A   B   C   D

a. In which of the tubes A to D will the seeds germinate? Explain your answers. (8)

b. The following results were obtained from an experiment to demonstrate that seed germination is affected by temperature. Seven lots of 100 seeds were kept on moist cotton wool and subjected to different temperatures. After 48 hours the number showing germination (at least 1 mm of root showing) were counted.

TEMPERATURE / °C	PERCENTAGE GERMINATION
5	3
15	22
25	55
35	79
45	52
55	19
65	2

i. Present the results in a graph. (5)
ii. What does the shape of the graph tell you about the control of germination in mustard seeds? (3)
iii. Why do gardeners sometimes scratch a hole in the seed coat before they plant the seeds? (2)

3. The diagram below shows a germinated seed.

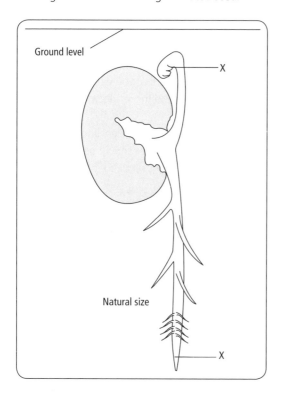

Ground level

X

Natural size

X

a. i. What food substance, stored in the seed, will be broken down into sugars? (1)
   ii. What food substance, stored in the seed, will be broken down into amino acids? (1)
b. Which class of chemicals will cause the stored food to break down into soluble compounds? (1)
c. Explain why amino acids are moved to the areas marked X in the diagram. (2)
d. Give **three** conditions that must have been present for this seed to germinate. (3)
e. i. Name fully the growth response being shown by the young shoot in the diagram. (2)
   ii. Explain the importance of this response to the plant. (1)

4. a. All seeds need oxygen, water and a suitable temperature to germinate. 22 °C is a suitable temperature for the germination of pea seeds. Light and dark conditions have no effect on pea seed germination. The diagram below shows an experiment on germination of pea seeds.

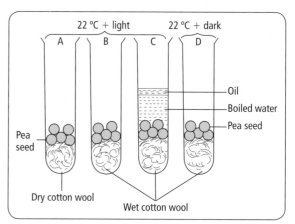

Complete the table below.

TUBE	WOULD SEEDS GERMINATE? (WRITE YES OR NO)
A	
B	
C	
D	

(4)

b. The next table shows how the dry mass of barley seedlings changed over the first 35 days after sowing.

TIME AFTER SOWING / DAYS	0	7	14	21	28	35
DRY MASS / g	4.0	2.8	2.8	4.4	9.6	17.8

i. Complete the plotting of data from the table on the graph below. The first two results have already been plotted for you.

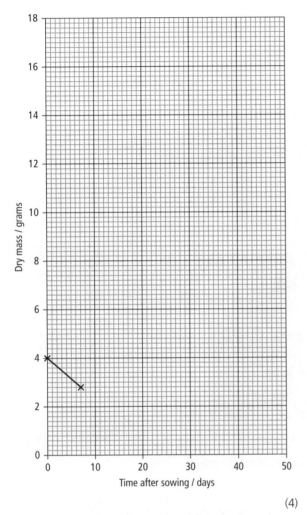

Time after sowing / days

(4)

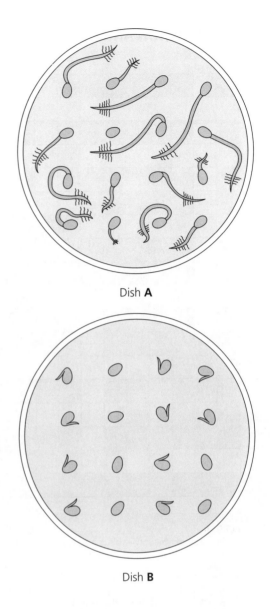

Dish **A**

Dish **B**

ii. How many days after sowing did the barley seedlings regain their original dry mass? (1)

iii. How many days after sowing did the barley seedlings **treble** their original dry mass? (1)

5. Two batches of tomato seeds were placed in dishes to germinate. Water was added to the seeds in dish **A**.

A 25% solution of the juice from a fresh tomato was added to the seeds in dish **B**.

The diagram shows the two batches after 4 days.

a. State **two** conditions, other than water, that are needed for the germination of the seeds. (1)

b. i. Describe briefly the results shown:
in dish **A**
in dish **B**

ii. Make an enlarged, labelled drawing of one of the seedlings from dish **A**. (5)

c. i. Suggest what has caused the difference between the two batches of seeds.

ii. Suggest how you would find out whether this factor, which affects the germination of tomato seeds, also affects the germination of other types of seeds. (4)

GERMINATION AND PLANT GROWTH: Crossword

ACROSS:
3   The fusion of male and female gametes
4   Enzyme necessary to break down starch stores in a seed
5   The part of the seed that develops into a young plant
7   A coat that prevents a seed from drying out
8   Part of the embryo that develops into a root
9   Contains the female gamete and develops into the seed
10 A seed leaf, forming a food store in a seed
13 Specialised leaves for sexual reproduction in plants
15 Part of the embryo that becomes the shoot
16 Catalysts affected by temperature in germinating seeds
17 Needed for cells to swell so the embryo can push through the seed coat

DOWN:
1   The development of a seed into a young plant
2   A sex cell, with the haploid number of chromosomes
6   Gas required for aerobic respiration in germinating seeds
11 A condition in which a seed may fail to germinate
12 Develops from the ovary following fertilisation
14 An enzyme that breaks down food stores in fatty seeds
15 Grain that contains the male gamete

DNA, genes and chromosomes

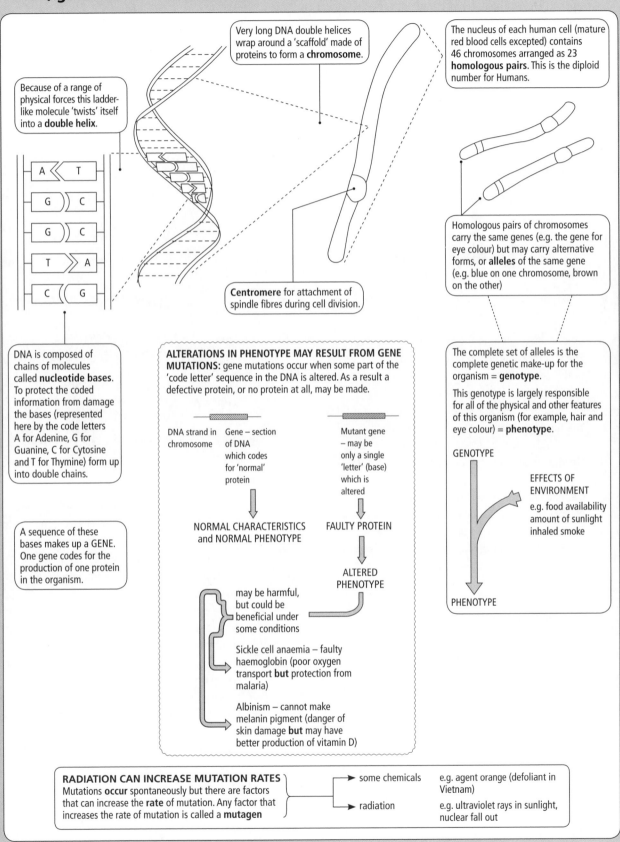

Because of a range of physical forces this ladder-like molecule 'twists' itself into a **double helix**.

Very long DNA double helices wrap around a 'scaffold' made of proteins to form a **chromosome**.

The nucleus of each human cell (mature red blood cells excepted) contains 46 chromosomes arranged as 23 **homologous pairs**. This is the diploid number for Humans.

Centromere for attachment of spindle fibres during cell division.

Homologous pairs of chromosomes carry the same genes (e.g. the gene for eye colour) but may carry alternative forms, or **alleles** of the same gene (e.g. blue on one chromosome, brown on the other)

DNA is composed of chains of molecules called **nucleotide bases**. To protect the coded information from damage the bases (represented here by the code letters A for Adenine, G for Guanine, C for Cytosine and T for Thymine) form up into double chains.

A sequence of these bases makes up a GENE. One gene codes for the production of one protein in the organism.

ALTERATIONS IN PHENOTYPE MAY RESULT FROM GENE MUTATIONS: gene mutations occur when some part of the 'code letter' sequence in the DNA is altered. As a result a defective protein, or no protein at all, may be made.

DNA strand in chromosome

Gene – section of DNA which codes for 'normal' protein

Mutant gene – may be only a single 'letter' (base) which is altered

NORMAL CHARACTERISTICS and NORMAL PHENOTYPE

FAULTY PROTEIN

ALTERED PHENOTYPE

may be harmful, but could be beneficial under some conditions

Sickle cell anaemia – faulty haemoglobin (poor oxygen transport **but** protection from malaria)

Albinism – cannot make melanin pigment (danger of skin damage **but** may have better production of vitamin D)

The complete set of alleles is the complete genetic make-up for the organism = **genotype**.

This genotype is largely responsible for all of the physical and other features of this organism (for example, hair and eye colour) = **phenotype**.

GENOTYPE

EFFECTS OF ENVIRONMENT e.g. food availability amount of sunlight inhaled smoke

PHENOTYPE

RADIATION CAN INCREASE MUTATION RATES Mutations **occur** spontaneously but there are factors that can increase the **rate** of mutation. Any factor that increases the rate of mutation is called a **mutagen**

some chemicals — e.g. agent orange (defoliant in Vietnam)

radiation — e.g. ultraviolet rays in sunlight, nuclear fall out

A — T
G — C
G — C
T — A
C — G

1. Chromosomes contain DNA. Each individual, with very few exceptions, has different DNA. The DNA can be broken up, and the pieces used to produce a 'genetic fingerprint'. The diagram shows the genetic fingerprints of blood found at a murder scene and from blood samples provided by five suspects.

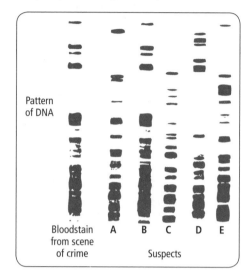

Fill in the missing words in the following sentences.

a. The chromosomes are found inside the .... of the cell. (1)

b. Genetic fingerprints must be made from the .... blood cells of a suspect. (1)

c. The murderer was most likely to be suspect .... (1)

d. Genetic fingerprints can only be identical if they come from .... (1)

e. Genetic fingerprinting can also be useful in conservation of endangered species. Scientists can examine the DNA fingerprint from a single hair of any animal. Suggest how this could be used to prevent dangerous inbreeding between rare animals kept in zoos. (2)

2. The diagram shows a section of a DNA molecule during replication.

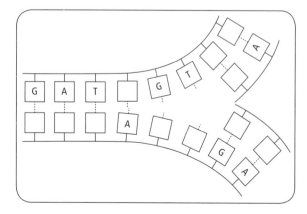

a. Copy the diagram, and complete it by writing in the correct letters in the empty boxes. (4)

b. Name the four molecules that these letters represent. (4)

c. DNA replication occurs during cell division. Why is replication important in the type of cell division called mitosis? (2)

d. What is the name of a section of DNA that codes for a particular protein? (1)

e. The table shows the codes for some of the amino acids found in human proteins.

ALANINE	GCC
CYSTEINE	ACA
GLYCINE	CCC
HISTIDINE	GTA
ISOLEUCINE	TAA
LYSINE	TTT
METHIONINE	TAC
PHENYLALANINE	AAA
PROLINE	CCG
TYROSINE	ATG
VALINE	CAT

The following is a sequence of bases in a section of DNA.

TAC – ATG – CCC – GCC – GTA

i. Write out the sequence of amino acids coded by this DNA if it is read from left to right. (2)

ii. Sometimes a base letter is lost, perhaps as a result of exposure to radiation. Write out the new code sequence **if the ninth letter (a C) is lost in this way**. (2)

iii. Write out the sequence of amino acids if this altered DNA sequence is used. (2)

iv. What name is given to the changes in DNA that can affect the production of proteins? (1)

v. Name one human disease that is caused by a faulty gene. (1)

vi. Explain why changes of this type can sometimes be an **advantage** to living organisms. (2)

3. The diagram shows some of the steps in the manufacture of proteins.

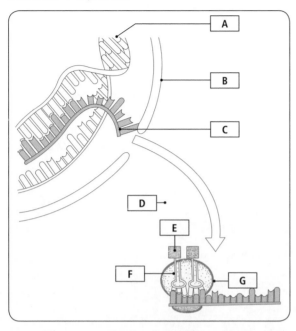

a. Match up the labels from this list with the spaces on the diagram. (7)
   TRANSFER RNA, CYTOPLASM, RIBOSOME, MESSENGER RNA, NUCLEAR MEMBRANE, DNA, AMINO ACID.
b. Name a protein found in
   i.   saliva
   ii. hair
   iii. red blood cells
   iv. stomach juices (4)

4. The sex of most animals is determined by their sex cells (gametes). It is these cells that combine to form a new organism. The diagram shows a male gamete from a human.

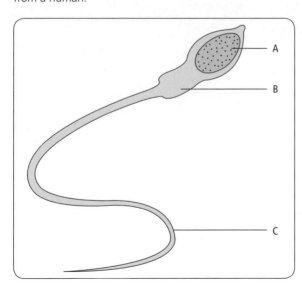

a. Name the type of cell division which produces gametes. (1)
b. State the major difference between a gamete and a normal cell. (1)
c. The part labelled A carries the instructions that will help to form the new organism. The instructions are carried as a chemical code. Name the chemical that carries these instructions. (1)
d. When a male and female gamete join together at fertilization a zygote is formed. The zygote divides many times to become an embryo, and eventually a baby organism. During this division the 'information chemical' is copied accurately. The diagram shows a section of this chemical during this copying. Draw a completed version of the diagram by adding the correct code letters. (2)

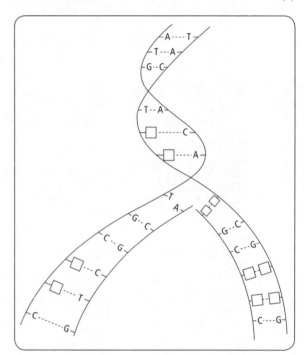

e. Sometimes the copying process goes wrong, and faulty information is made. What is the name given to the process causing 'faulty' copying? (1)
f. Name one environmental factor that might increase the risk of this faulty copying. (1)
g. Name one human disease caused by a fault of this type. (1)

5. a. The diagram shows the amount of DNA in a cell during cell division.

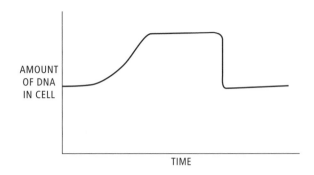

AMOUNT
OF DNA
IN CELL

TIME

i. Name the type of cell division occurring in the cell. Explain how you arrived at your answer. (2)
ii. What is the function of DNA? (2)

b. The DNA content of a cell may be altered as a result of radiation. What name is given to this kind of change in a cell's DNA? (1)

c. The DNA content of a cell may also be changed deliberately, by genetic engineering. Use the following terms to explain this technique. BACTERIUM, CLONE, PLASMID, RESTRICTION ENZYME, LIGASE (5)

SUMMARY: Matching pairs on DNA and characteristics

Match up the words in the first column with their definitions in the second column.

A	Protein	1	DNA wound onto a protein scaffold
B	DNA	2	A single code letter in a DNA chain
C	Chromosome	3	Biological molecule responsible for a characteristic of an organism
D	Gene	4	Two chromosomes with the same genes in the same positions
E	Phenotype	5	This molecule is faulty in sufferers from sickle cell anaemia
F	Genotype	6	Ultraviolet, for example, can cause mutations
G	Nucleotide	7	The genetic material of all animals and plants
H	Homologous pair	8	An individual with a non-pigmented skin
I	Mutation	9	Alternative forms of the same gene
J	Haemoglobin	10	The total of the physical and other features of an organism
K	Albino	11	The arrangement of the DNA molecule
L	Radiation	12	A change in the type or amount of DNA
M	Alleles	13	A sequence of nucleotide bases coding for a single protein
N	Double helix	14	The complete set of and alleles in a cell

DNA, GENES AND PROTEINS: Crossword

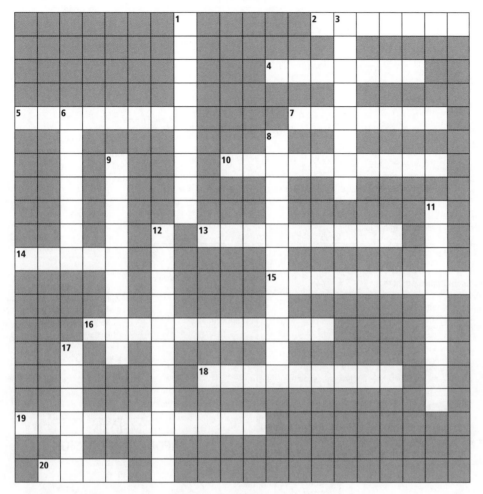

ACROSS:
2 Protein essential to the function of saliva
4 Type of molecule responsible for a characteristic in a cell or organism
5 The total of all the genetic material in a nucleus
7 Some factor which can change DNA
10 A long strand of genes
13 Type of RNA which carries genetic information from the nucleus to the cytoplasm
14 The 'letters' that make up the code in DNA
15 Sub-unit of a protein
16 Technical name for the copying of DNA
18 The total of all the characteristics of an organism
19 Important protein for oxygen transport
20 A section of DNA that codes for a single protein

DOWN:
1 This type of cell has DNA but not inside a nucleus
3 A change to the type or amount of DNA
6 where to look for DNA in a plant, animal or fungal cell
8 Process that 'read' the message from the nucleus and produces a protein
9 The organelles in the cytoplasm where proteins are produced
11 Ultraviolet, for example – an example of 7 across
12 The arrangement of the DNA chains
17 An alternative version of 20 across

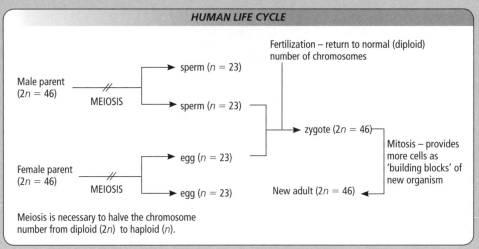

HUMAN LIFE CYCLE

Fertilization – return to normal (diploid) number of chromosomes

Male parent (2*n* = 46) — MEIOSIS → sperm (*n* = 23), sperm (*n* = 23)

Female parent (2*n* = 46) — MEIOSIS → egg (*n* = 23), egg (*n* = 23)

→ zygote (2*n* = 46)

Mitosis – provides more cells as 'building blocks' of new organism

New adult (2*n* = 46)

Meiosis is necessary to halve the chromosome number from diploid (2*n*) to haploid (*n*).

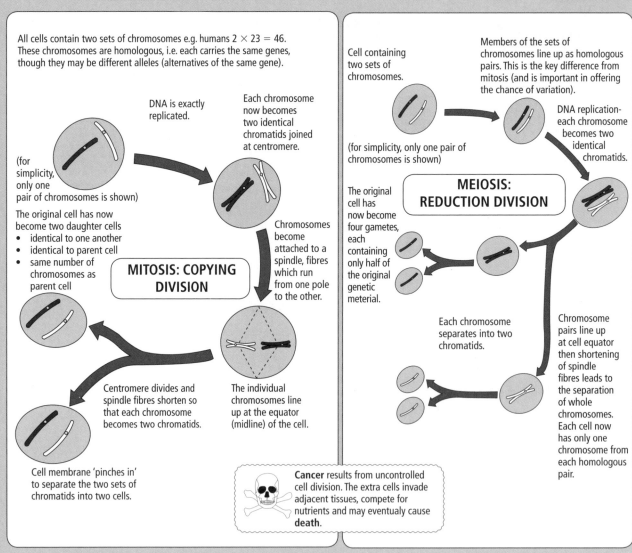

All cells contain two sets of chromosomes e.g. humans 2 × 23 = 46. These chromosomes are homologous, i.e. each carries the same genes, though they may be different alleles (alternatives of the same gene).

DNA is exactly replicated.

Each chromosome now becomes two identical chromatids joined at centromere.

(for simplicity, only one pair of chromosomes is shown)

The original cell has now become two daughter cells
• identical to one another
• identical to parent cell
• same number of chromosomes as parent cell

MITOSIS: COPYING DIVISION

Chromosomes become attached to a spindle, fibres which run from one pole to the other.

Centromere divides and spindle fibres shorten so that each chromosome becomes two chromatids.

The individual chromosomes line up at the equator (midline) of the cell.

Cell membrane 'pinches in' to separate the two sets of chromatids into two cells.

Cell containing two sets of chromosomes.

Members of the sets of chromosomes line up as homologous pairs. This is the key difference from mitosis (and is important in offering the chance of variation).

(for simplicity, only one pair of chromosomes is shown)

DNA replication- each chromosome becomes two identical chromatids.

The original cell has now become four gametes, each containing only half of the original genetic material.

MEIOSIS: REDUCTION DIVISION

Each chromosome separates into two chromatids.

Chromosome pairs line up at cell equator then shortening of spindle fibres leads to the separation of whole chromosomes. Each cell now has only one chromosome from each homologous pair.

Cancer results from uncontrolled cell division. The extra cells invade adjacent tissues, compete for nutrients and may eventualy cause **death**.

1. a. What is a chromosome?
   b. How many chromosomes are there in
      i. A skin cell from a human male?
      ii. A human egg cell?
      iii. A red blood cell from a human female?
      iv. A red blood cell from a human male?   (4 × 1)
   c. Name the process in which haploid cells are formed from diploid cells.   (1)
   d. Name the process which provides new cells for the growth of a young mammal.   (1)

2. This diagram represents the life cycle of a mammal.

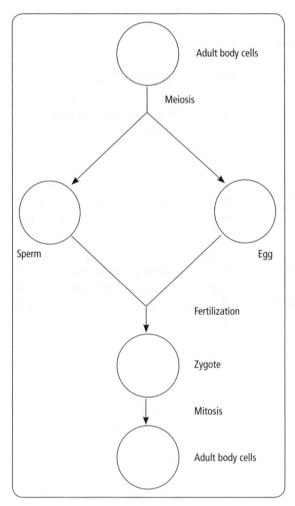

   a. Copy and complete the diagram by writing in the circles the numbers of chromosomes found in the nuclei of these cells if the mammal concerned was a human.   (5)
   b. In what way is this diagram misleading?   (1)
   c. Where, exactly, does meiosis occur in a mammal? (1)
   d. Why is it necessary for gametes to be formed by meiosis?   (2)

3. It is possible to observe the process of mitosis in cells that are actively dividing. The tips of roots and shoots have many dividing cells when a plant is increasing in length.

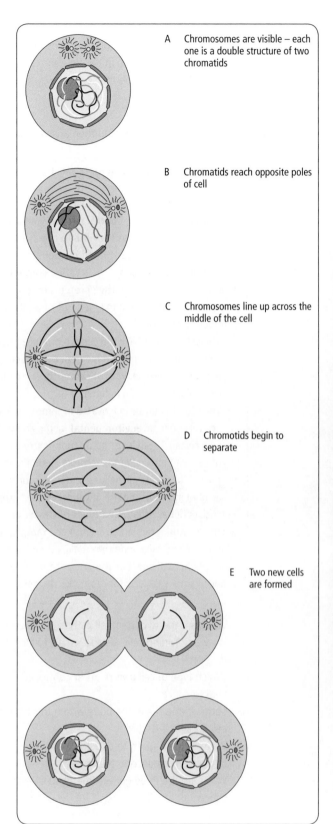

A Chromosomes are visible – each one is a double structure of two chromatids

B Chromatids reach opposite poles of cell

C Chromosomes line up across the middle of the cell

D Chromotids begin to separate

E Two new cells are formed

A student was able to cut off the top 5 mm of a plant shoot. She then softened it by warming it in a dilute solution of hydrochloric acid, and squashed the tip onto a microscope slide. The squashed cells were treated with a red dye, and then observed under a light microscope.
   a. Why did she soften the tip?   (1)
   b. Why did she squash the tip?   (1)

c. Why is red dye added to the preparation? (1)
d. The student took some photographs of the dividing cells. These are shown in the figure on the previous page. What would be the sequence of these cells as they divided? (4)

e. Make a labelled drawing of one chromosome from figure A. (2)
f. How many chromatids are present in figure B? (1)
g. How many chromosomes would be present in a body cell from this organism? (1)

4. Read the following article, and then answer the questions that follow it.

WHAT CAUSES CANCERS?

The rate of cell division by mitosis is normally very strictly controlled. The control normally involves cells touching one another as they fill up the available space – once the space is properly filled the cells 'switch off' their process of division. This is called **contact inhibition** and involves special sensitive proteins on the cell surface membrane. Cell division may get out of control for a number of reasons:

- The cell may not have the gene to produce the correct cell surface protein – this is a **genetic tendency**, and explains why some types of cancer (breast cancer is an example) tend to run in families.
- The protein may be inactivated in some way. This usually occurs due to an **environmental** factor, and explains why certain lifestyles or occupations carry a higher risk of cancer. Smoking of tobacco, for example, increases the risk of cancer of the lung and heavy consumption of alcohol increases the risk of cancer of the oesophagus.

Scientists believe that most types of cancer only develop when several factors are present. For example, a person might be carrying a gene that makes their lung cells more likely to divide, but the presence of chemicals in tobacco smoke is necessary to make the gene run out of control. Any factor in the environment that causes cancer to develop is called a **carcinogen** (literally a 'cancer maker') – the involvement of external factors in the development of cancer of the lung, for example, has been very widely studied.

THE TREATMENT OF CANCER

If a cancer is detected early enough it can often be treated. Hospitals run **screening programmes** to try to detect some cancers, such as breast cancer, cervical cancer and lung cancer, while they are still treatable. Doctors urge their patients to visit the surgery if they have any concern about their health, especially sudden weight loss or unexplained bleeding. Treatment always involves exerting some control of the cancer cells again, either by surgery, chemotherapy or radiotherapy.

'Cancer' is actually a very general term. There are many forms of cancer, and many tissues may develop tumours. Since cancer results from cell division that has gone out of control, the disease is most likely to occur in tissues that are normally dividing quite frequently. For this reason cancer is particularly common in the sites in the body where mitosis is occurring.

a. What are the two 'general' causes of cancer? (2)
b. How do dividing cells 'decide' when to stop their division? (1)
c. Which type of cell division, if uncontrolled, can lead to cancer? (1)
d. Suggest one site in the body where this type of cell division is occurring. (1)
e. What name is given to a factor that causes cells to divide in an uncontrolled way? (1)
f. What are the three general methods of treating cancer? (3)

5. The figure below shows stages in the formation of a human fetus.

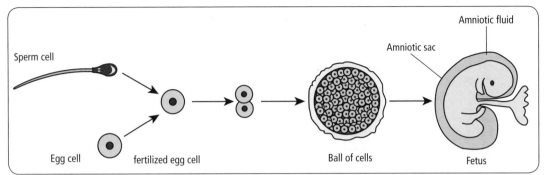

a. i.   Name the process of cell division that results in the formation of sperm cells.   (1)

  ii. State one way in which the sperm cell is different from cells in the developing fetus.   (1)

  iii. State the term used to describe the fertilized egg cell.   (1)

  iv. Explain what determines that a fertilized egg cell develops into a girl rather than a boy.   (1)

b. State where each of the following is produced.
  i.   the egg cell
  ii. the fertilized egg
  iii. the fetus   (3)

CIE 0610 June '05 Paper 3 Q5

SUMMARY: Matching pairs on cell division

Match up the words in the left hand column with the definitions in the right hand column.

	TERM		DESCRIPTION
A	Mitosis	1	The number of pairs of chromosomes in a normal body cell
B	Meiosis	2	A copied chromosome
C	Fertilization	3	Structure that holds the chromosome to the spindle as cell division proceeds
D	Zygote	4	A sex cell – the product of meiosis
E	Haploid	5	Reduction division, which produces haploid daughter cells from a diploid parent cell
F	Diploid	6	The diploid product of fertilization
G	Chromatid	7	Copying division, essential for growth, repair and asexual reproduction
H	Spindle	8	The fusion of gametes as part of sexual reproduction
I	Centromere	9	The total number of chromosomes in a normal body cell
J	Gamete	10	Framework for moving chromosomes or chromatids to the poles during cell division

Cystic fibrosis in humans is an example of monohybrid inheritance

Cystic fibrosis (cf) results from an imbalance of chloride ions across the membranes of cells lining some of the main passageways of the body, causing dangerous build up of mucus.

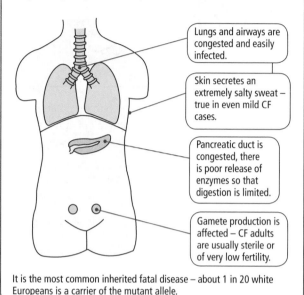

Lungs and airways are congested and easily infected.

Skin secretes an extremely salty sweat – true in even mild CF cases.

Pancreatic duct is congested, there is poor release of enzymes so that digestion is limited.

Gamete production is affected – CF adults are usually sterile or of very low fertility.

It is the most common inherited fatal disease – about 1 in 20 white Europeans is a carrier of the mutant allele.

SOME IMPORTANT DEFINITIONS

HOMOZYGOUS – having two identical alleles for a particular gene e.g. NN or nn
HETEROZYGOUS – having two different alleles for a particular gene e.g. Nn
DOMINANT – an allele which is expressed if it is present. Usually given a capital letter e.g. N
RECESSIVE – an allele which is expressed only when there is no dominant allele of the gene present. Usually given a small letter e.g. n

If both parents are carriers (i.e. heterozygous)

PARENTAL GENERATION   Nn   ×   Nn

GAMETES   Ⓝ   ⓝ   Ⓝ   ⓝ

A Punnett square can be used to predict the possible combinations of alleles in the zygote

Gametes from father ♂	Ⓝ	ⓝ
Gametes from mother ♀ Ⓝ	NN	Nn
ⓝ	Nn	nn

Phenotype: no cystic fibrosis
Genotype: normal

Phenotype: no cystic fibrosis
Genotype: carrier of mutant allele

Phenotype: has cystic fibrosis
Genotype: homozygous for mutant allele

3 no cystic fibrosis :   1 with cystic fibrosis

If one parent's genotype is homozygous normal, the other parent's is homozygous mutant, then theoretically:

The homozygous normal parent is represented as NN, the parent with homozygous mutant genotype as nn, since in this case normal is dominant to mutant.

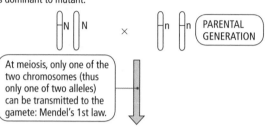

N   N   ×   n   n   PARENTAL GENERATION

At meiosis, only one of the two chromosomes (thus only one of two alleles) can be transmitted to the gamete: Mendel's 1st law.

Ⓝ   ⓝ   GAMETES

At fertilization, fusion of gametes to form a zygote restores the diploid number.

1st FILIAL GENERATION

Nn

The allele that 'shows up' in this heterozygote is DOMINANT, the other allele (which remains 'hidden') is RECESSIVE

This individual is **genotypically** heterozygous, but **phenotypically** normal i.e. does not have the disease but is a **carrier** of the mutant allele for cystic fibrosis.

SOLVING INHERITANCE PROBLEMS

THE KEY! The only genotype you can be sure of is the homozygous recessive, e.g. CF sufferer **must** be nn…

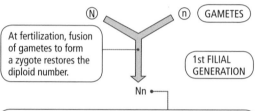

…so non-CF parents **must** be Nn (since each must hand on one n allele to the offspring).

Nn   Nn

CF
nn

N.B. The 3:1 ratio is only approximate unless the number of offspring is very large (unlikely in humans), because

1 alleles may not be distributed between viable gametes in equal numbers
2 fusion of gametes is completely random. It is a matter of chance whether one male gamete fuses with a particular female gamete.

Sex linkage and the inheritance of sex

A KARYOTYPE is obtained by rearranging photographs of stained chromosomes observed during mitosis. Such a karyotype indicates that

① the chromosomes are arranged in **homologous pairs**. In humans there are 23 pairs and we say that the **diploid number** is 46 ($2n = 46 = 2 \times 23$).

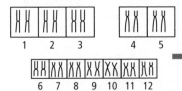

```
1   2   3      4   5
6   7   8   9   10  11  12
13  14  15  16  17  18  19  20
21  22
        23(x)
```

② whereas females have 22 pairs + XX in the karyotype, males have 22 pairs + XY, i.e. the 23rd 'pair' would **not** be two copies of the X chromosome.

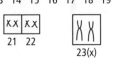

♀
(XX)

♂
(XY)

The Y chromosome is so small that there is little room for any genes other than those responsible for 'maleness', but the X chromosome can carry some genes as well as those for 'femaleness' – these additional genes are **X-linked** (usually described as **sex-linked**).

Another important X-linked condition is **red-green colour blindness**. Many more males than females cannot distinguish red from green.

ALTERATIONS IN PHENOTYPE MAY RESULT FROM CHROMOSOME MUTATION
A mutation is an alteration in the DNA content of a cell. Chromosome mutations occur when the processes of cell division fail to work with complete accuracy. For example, during meiosis

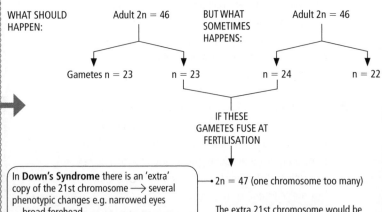

WHAT SHOULD HAPPEN:

Adult $2n = 46$

BUT WHAT SOMETIMES HAPPENS:

Adult $2n = 46$

Gametes $n = 23$ $n = 23$ $n = 24$ $n = 22$

IF THESE GAMETES FUSE AT FERTILISATION

In **Down's Syndrome** there is an 'extra' copy of the 21st chromosome ⟶ several phenotypic changes e.g. narrowed eyes
  broad forehead
  heart abnormalities

$2n = 47$ (one chromosome too many)

The extra 21st chromosome would be seen in the child's karyotype

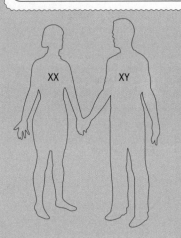

XX   XY

CODOMINANCE

Some genes have more than two alleles. For example, the gene controlling the human ABO blood groups has three alleles, given the symbols I^A, I^B and I^O. Neither of the I^A and I^B alleles is dominant to the other, although they are both dominant to I^O. This is called **codominance**. It results in an extra phenotype when both alleles are present together. The genotypes and phenotypes are shown in the table.

Genotype	Phenotype
$I^A\,I^A$ or $I^A\,I^O$	Blood group A
$I^B\,I^B$ or $I^B\,I^O$	Blood group B
$I^A\,I^B$	Blood group AB
$I^O\,I^O$	Blood group O

The human blood groups are easily detected by a simple test on a blood sample.

INHERITANCE OF SEX is a special form of monohybrid inheritance

♂ (XY)   ♀ (XX)

GAMETES (X) (Y)   (X) (X)

F_1 generation: sex of offspring can be determined from a Punnett square

♀ GAMETES \ ♂ GAMETES	X	Y
X	XX (female)	XY (male)
X	XX (female)	XY (male)

- theoretically there should be 1:1 ratio of male : female
- the male gamete determines the sex of the offspring

This can be interesting when a woman can't be sure of the father of a child. For example, could a type O man be the father of a type AB child?

♂ gametes	I^O	I^O
♀ gametes: no ♀ could provide both I^A and I^B		Child $I^A\ I^B$

NO! This man could NOT be the father!

1. Choose words from the following list to match the definitions below.

    GENOTYPE, ALLELE, MENDEL, DARWIN, GENE, PHENOTYPE, HETEROZYGOTE, DOMINANT, HOMOZYGOTE

    a. The external appearance of an organism.
    b. Alternative form of a gene.
    c. Studied genetics in garden peas.
    d. Section of DNA coding for a single characteristic.
    e. The set of genes in the nucleus of any individual.
    f. A nucleus carrying both alternative alleles of a gene.
    g. An allele that determines the appearance of a heterozygote. (7 × 1)

2. Cystic fibrosis is one of the most common inherited diseases in humans. The disease is caused by the inheritance of a recessive allele – about 1 in 20 white-skinned individuals are heterozygous for this allele. Heterozygotes do not show symptoms of the disease, but are 'carriers' of the allele.
    a. Copy and complete the following genetic diagram to show the possible inheritance of cystic fibrosis by children from two 'carrier' parents. (Let R = normal allele; let r = allele for cystic fibrosis) (7)

    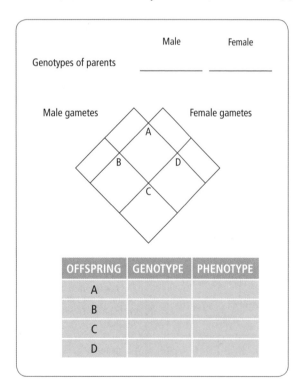

OFFSPRING	GENOTYPE	PHENOTYPE
A		
B		
C		
D		

    b. Two 'carrier' parents have two children, neither of whom show any symptoms of cystic fibrosis. What is the chance that a third child will have cystic fibrosis? (2)
    c. It is possible to use a gene probe to detect the cystic fibrosis allele – the probe is able to bind onto this allele in a sample of a person's DNA. The DNA is collected from blood. Which type of blood cell would supply the DNA for the test? Explain your answer. (2)

3. Haemophilia is a sex-linked inherited characteristic. The allele (n) for haemophilia is recessive to the allele (N) for normal blood clotting. The gene responsible for blood clotting is carried only on the X chromosome.

    The phenotype of an organism is the appearance or characteristic that results from the inheritance of a particular combination of alleles.
    a. Describe the phenotypes of individuals with the following alleles. (4)

    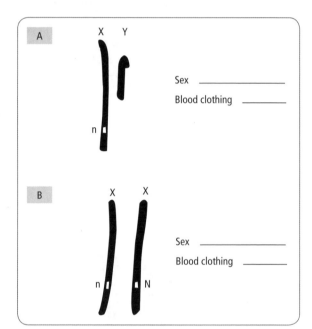

    b. Use genetic symbols to explain how two non-haemophiliac parents could have a haemophiliac son. (4)
    c. Explain
       i. why haemophiliac females are less common than haemophiliac males
       ii. why haemophilia is especially dangerous for girls. (2)

4. Sickle cell anaemia is a disease in which people produce abnormal haemoglobin in their red blood cells. This abnormal haemoglobin makes the cells take on a sickle shape, especially if the person exercises and produces lactic acid. The sickle cells do not carry oxygen as well as normal cells do.

    Let N = the allele for 'normal' haemoglobin; let S = the allele for 'sickle' haemoglobin.
    a. Copy and complete the genetic diagram to show the results of a cross between two 'carrier' parents. (4)

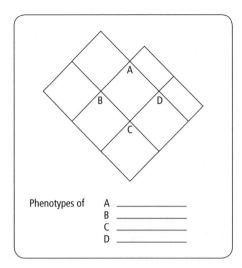

Phenotypes of   A _____
                 B _____
                 C _____
                 D _____

b. Explain why a person with sickle cell anaemia would often feel tired. (1)

c. People who are carriers of this condition are resistant to a blood-borne disease transmitted by mosquitoes. What is the name of this disease? (1)

d. Name the process in which a 'normal' allele might be converted to a 'sickle' allele. (1)

e. Name **one** environmental factor that might cause a change of this type. (1)

5. Cystic fibrosis is a serious condition in humans. Many of the body tubes become blocked with a very sticky mucus. Bacteria invade the mucus and cause serious infections, and tubes and passageways blocked by the mucus do not allow the passage of important juices. Organs most affected by the condition are the lungs, pancreas and reproductive organs. The disease is inherited, and involves a single gene with two alleles.

a. Two parents who showed no symptoms of the condition had two children. The second child had cystic fibrosis although the other child was healthy.
  i. Is the allele that causes cystic fibrosis dominant or recessive? Explain your answer. (2)
  ii. The parents decide to have another child. What is the chance that the third child will have cystic fibrosis? Use a genetic diagram to explain your answer. (4)

b. The condition is treated by giving patients antibiotics and capsules of enzymes. Explain why this treatment is effective. (2)

c. Read this extract from a scientific journal.

Medical researchers have developed a method for correcting cystic fibrosis. The treatment involves introducing a normal allele of the gene for a salt-transporting protein into the lungs of patients suffering from this condition. The normal gene is inserted into the coat of a virus similar to the one that causes the common cold. The virus's own genetic ma terial is removed. The patient inhales the genetically engineered virus, the lung cells are invaded by the virus and stop producing the sticky mucus.

  i. Draw a series of diagrams to explain how the virus could be genetically engineered to contain the 'normal' gene. (4)
  ii. Why are some people very much against the use of genetically engineered organisms? (3)

REVISION SUMMARY: Fill in the gaps

Use terms from the following list to complete the paragraphs below. You may use each term once, more than once or not at all.

GENES, NUCLEUS, DARWIN, MENDEL, TALL, DWARF, 3:1, DNA, 1:1, CHROMOSOMES, STAMENS, INSECTS, GARDEN PEA, STIGMA, HETEROZYGOUS, RECESSIVE, DOMINANT, HOMOZYGOUS, HEIGHT

An Austrian monk called .... carried out a series of important genetic investigations on plants of the .... To control the inheritance of characteristics he removed the .... from some plants and dusted their pollen onto the .... of other plants of the same species. He then covered these flowers in muslin bags so that they could not be pollinated by ....

In one experiment he crossed .... plants with dwarf plants to investigate the inheritance of .... in this species. When he used pure-breeding (........................) parent plants, all of the offspring were .... – this is the .... characteristic. When these offspring were self-fertilized they produced tall and dwarf plants in a ratio of ....

Modern work in biochemistry was unknown to ...., and he knew nothing of the genetics familiar to us. For example, he was unaware that characteristics are controlled by short sections of .... carried as .... on .... He did, however, have the use of a microscope and did appreciate that the 'factors of inheritance' were carried in the .... of the cell. (16)

Variation and natural selection may lead to evolution of species

Environmental resistance describes the factors, e.g. food availability, that limit the growth of populations. Animals and plants that are best **adapted** to their environment suffer from less environmental resistance. **Adaptations** are often structural, but they can also be biochemical or behavioural.

These adaptations arise because there is **variation** between the different members of a population. Charles Darwin studied many examples of these adaptations, and published his conclusions in *The Origin of Species by means of Natural Selection*. His observations can be summarized under a number of headings.

Over-production: All organisms produce more offspring than can possibly survive, but populations remain relatively stable.

e.g. a female peppered moth may lay 500 eggs, but the moth population does not increase by 25 000 %!

Struggle for existence: Organisms experience environmental resistance, i.e. they compete for the limited resources within the environment.

e.g. several moths may try to feed on the same nectar-producing flower.

Variation: Within the population there may be some characteristics that make the organisms that possess them more suited for this severe competition.

e.g. moths might be stronger fliers, have better feeding mouthparts, be better camouflaged while resting or be less affected by rain.

Survival of the fittest: Individuals that are most successful in the struggle for existence (i.e. are the best suited/adapted to their environment) will survive more easily than those without these advantages.

e.g. peppered moths: dark coloured moths resting on soot-covered tree trunks will be less likely to be captured by predators.

Advantageous characteristics are passed on to offspring: The well adapted individuals breed more successfully than those that are less well adapted – they pass on their genes to the next generation. This process is called **natural selection**.

e.g. dark coloured moth parents will produce dark coloured moth offspring.

HOW GENETIC VARIATION MAY ARISE

- both chromosomal and gene **mutation** provide **raw material for variation**
- meiosis and sexual reproduction then **re-arrange this raw material** to provide **many new genotypes**, by (a) crossing over, (b) independent assortment and (c) fertilization

TYPES OF VARIATION

1 DISCONTINUOUS: caused by GENES only

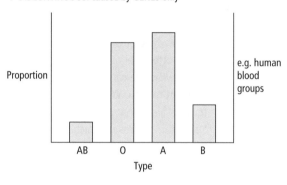

e.g. human blood groups

2 CONTINUOUS: caused by GENES and EFFECTS OF ENVIRONMENT

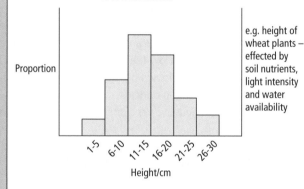

e.g. height of wheat plants – effected by soil nutrients, light intensity and water availability

SUMMARY:

VARIATION $\xrightarrow{\text{NATURAL SELECTION}}$ ADAPTATION $\xrightarrow{\text{MORE NATURAL SELECTION}}$ NEW SPECIES

1. The table shows some examples of variation in living organisms. For each of the examples, indicate whether the variation results from **artificial** (A) or from **natural** (N) selection.

NUMBER	VARIATION OBSERVED	ARTIFICIAL OR NATURAL
1	Cattle bred for high milk yield	
2	Development of wheat strains resistant to infection by fungi	
3	Sheep bred for thick wool	
4	The development of thick fur by polar bears	
5	Bacteria developing antibiotic resistance	
6	Banding on snail shells offers camouflage from song thrushes	
7	Birch trees growing on mine spoil heaps with a high lead content	
8	Members of the *Brassica* (cabbage) group chosen as sprouts because of many small side buds	

2. A group of students carried out an investigation into the diameter of the bases of limpet shells from the same rocky shore. They presented their results in the two histograms shown below:

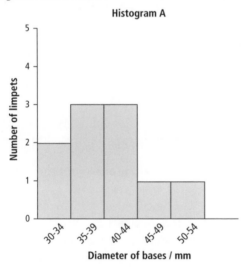

Histogram A

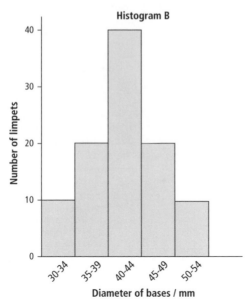

Histogram B

a. How many limpet shells were measured to produce the data shown in the two histograms? (2)

b. Which of the two histograms, A or B, would give the more reliable mean diameter of the base of the limpet shells? Explain your answer. (2)

The limpet uses its muscular foot to hold onto rocks. The diameter of the base of the shell is a very good measure of the size of the foot. The students carried out a further investigation into base diameter of limpets from two different parts of the shore.

	AREA 1	AREA 2
Mean diameter of shell bases / mm	39	47

c. Which of the two areas, 1 or 2, had the biggest waves? Explain your answer. (2)

d. What type of variation is shown by the diameter of the bases of the limpet shells? (1)

e. Give an example of the same type of variation in humans. (1)

3. Snails of one species vary as shown.

A scientist made the following observations on a group of snails.

1. A large number of snails are born.

2. Some snails have no bands on their shells. Others have bands on their shells.

3. Snails without bands on their shells tend to produce young without bands on their shells.
4. In hot conditions, snails with bands on their shells are more likely to die of heat shock.

The table gives **three** statements about the theory of evolution by natural selection.

a. Write the correct numbers in the table so that the observations made by the scientist match the following statements about evolution.

STATEMENT	NUMBER OF MATCHING OBSERVATION
Some variations are inherited	
All populations show variation	
Natural populations over-produce	

(3)

b. A scientist did an investigation in which he used a special paint to mark snail shells.

The paint was sensitive to daylight.

He put a small spot of this paint on the top of the snails' shell.

He used 200 snails with banded shells, and 200 snails with unbanded shells.

He put the snails into ten cages in areas planted with grass and nettles.

Sixty days later he looked at how much the paint had faded.

The table shows his results.

The higher the figure, the more faded the paint is.

	BANDED SNAIL SHELLS	UNBANDED SNAIL SHELLS
Average fading of paint spot on shell	4.55	3.85

i. Why did he not have to worry about the paint spots making the snails easy to see by predators? (1)
ii. How was he able to work out which type of snail stayed longer in the sunlight. (1)
iii. Why would putting a spot of paint in the same place on each snail's shell help make it a fair test? (1)
iv. Suggest why dark banded snails have an advantage over unbanded snails in cold and shady places. (2)

4. The diagram shows a plant that is well adapted to life in a particular environment.

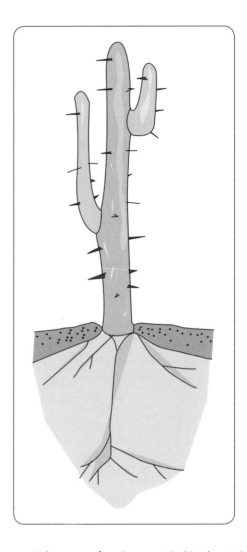

a. What type of environment is this plant suited to? Explain your answer. (2)
b. Describe three features that suit the plant to its environment. (3)
c. Match the letters and the numbers to explain how the mammal shown below is well adapted to its environment. (6)

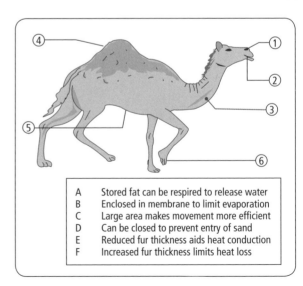

A	Stored fat can be respired to release water
B	Enclosed in membrane to limit evaporation
C	Large area makes movement more efficient
D	Can be closed to prevent entry of sand
E	Reduced fur thickness aids heat conduction
F	Increased fur thickness limits heat loss

REVISION SUMMARY: Fill in the gaps

Complete the following paragraphs. Use words and phrases from the following list – you may use each word once, more than once or not at all.

GENES, FERTILIZATION, EFFECTS OF ENVIRONMENT, PHENOTYPE, MUTATION, DISCONTINUOUS, CONTINUOUS, NATURAL SELECTION, EVOLUTION, CROSSING OVER, BLOOD GROUPING, ENVIRONMENTAL, ATMOSPHERIC, INDEPENDENT ASSORTMENT, GENOTYPE, BODY MASS, NUTRIENTS.

There are two kinds of variation – the first is ...., which shows clear cut separation between groups showing this variation (........................, for example). The second is ...., in which there are many intermediate forms between the extremes of the characteristic. A clear example of this second type is ....

.... is the result of .... alone, while .... is also affected by .... factors. The sum of the genes that an organism contains is called its .... and the total of all its observable and measurable characteristics is called the .... The two are related in a simple equation .... equals .... plus ....

Variation that can be inherited results from ...., which provides the raw material that can be rearranged by .... and .... during meiosis, and the process of .... when male and female gametes fuse.

Variation provides the raw material for .... Organisms may gain an advantage in the struggle for existence, and .... can 'pick out' the organisms with such an advantage. These organisms may then reproduce and pass on their .... to their offspring.

(20)

Chapter 33:
Ecology and ecosystems

Habitats, communities and populations

In natural habitats, many species of plants and animals live and interact together.

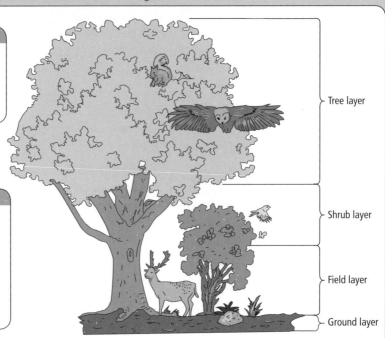

HABITAT

The place where an animal or plant lives is its **habitat**. A habitat provides food, shelter and space for all the organisms within it to live and reproduce.

This is a woodland habitat.

COMMUNITIES

Several species of plants and animals live and interact in this habitat. They form a **community**.

Each organism is adapted for life within its part of the habitat. For example, squirrels are well adapted for climbing and eating nuts in the tree layer. Woodland flowers are well adapted for growing and flowering under tall, deciduous trees.

- Tree layer
- Shrub layer
- Field layer
- Ground layer

ECOSYSTEMS

An ecosystem is a unit containing all of the organisms (in the community) and their environment in a given area. An example would be a woodland.

POPULATIONS

All the individuals of the same species in a habitat make up a **population**. Under ideal conditions plants and animals reproduce quickly and the population grows (1). However, the habitat can only produce a certain amount of food. This will limit the size of the population (2). If the population damages the habitat too much, the numbers may start to fall (3).

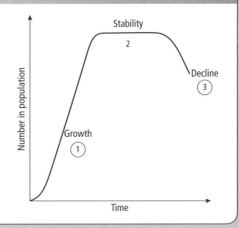

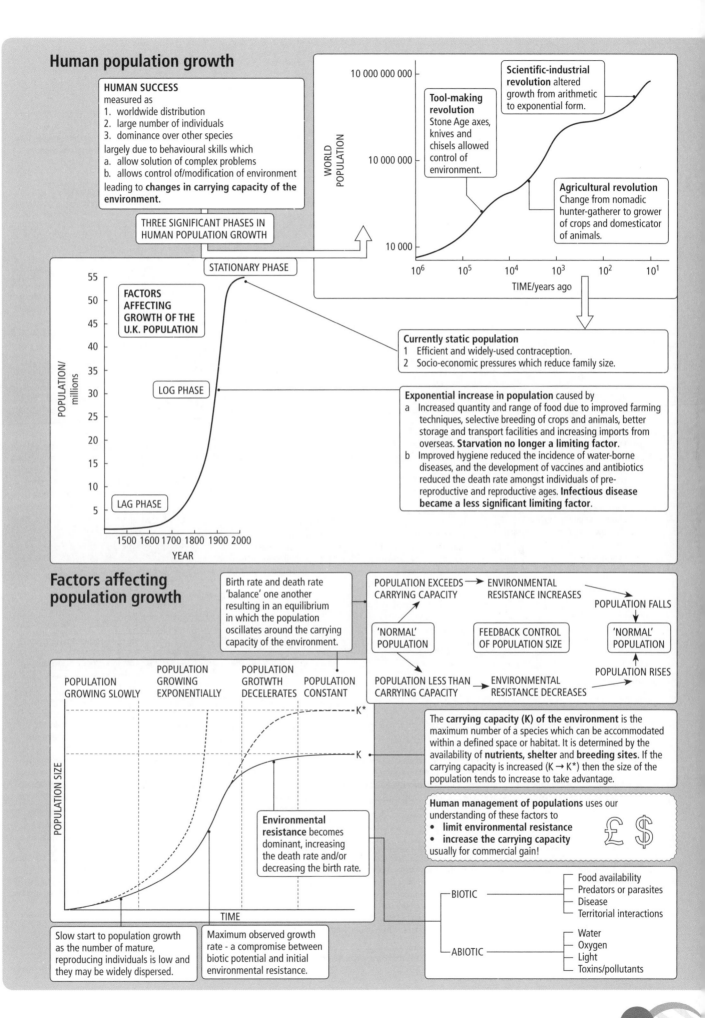

Ecological pyramids

Represent numerical relationships between successive trophic levels in a community.

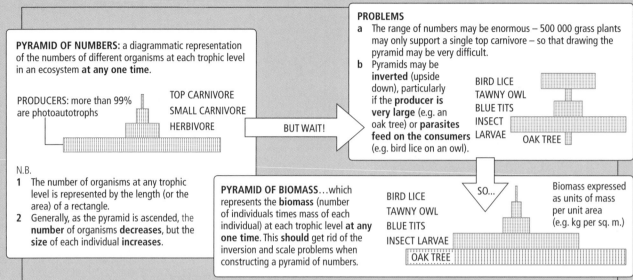

PYRAMID OF NUMBERS: a diagrammatic representation of the numbers of different organisms at each trophic level in an ecosystem **at any one time**.

PRODUCERS: more than 99% are photoautotrophs

TOP CARNIVORE
SMALL CARNIVORE
HERBIVORE

N.B.
1 The number of organisms at any trophic level is represented by the length (or the area) of a rectangle.
2 Generally, as the pyramid is ascended, the **number** of organisms **decreases**, but the **size** of each individual **increases**.

BUT WAIT!

PROBLEMS
a The range of numbers may be enormous – 500 000 grass plants may only support a single top carnivore – so that drawing the pyramid may be very difficult.
b Pyramids may be **inverted** (upside down), particularly if the **producer is very large** (e.g. an oak tree) or **parasites feed on the consumers** (e.g. bird lice on an owl).

BIRD LICE
TAWNY OWL
BLUE TITS
INSECT LARVAE
OAK TREE

SO...

PYRAMID OF BIOMASS…which represents the **biomass** (number of individuals times mass of each individual) at each trophic level **at any one time**. This **should** get rid of the inversion and scale problems when constructing a pyramid of numbers.

BIRD LICE
TAWNY OWL
BLUE TITS
INSECT LARVAE
OAK TREE

Biomass expressed as units of mass per unit area (e.g. kg per sq. m.)

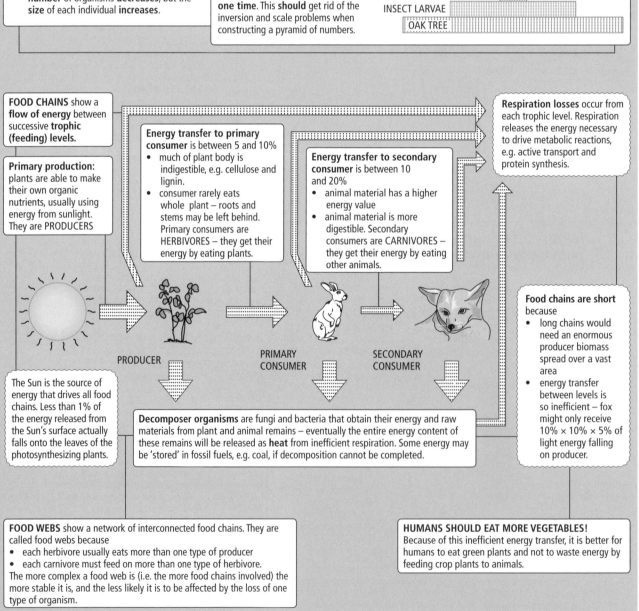

FOOD CHAINS show a **flow of energy** between successive **trophic (feeding) levels**.

Primary production: plants are able to make their own organic nutrients, usually using energy from sunlight. They are PRODUCERS

Energy transfer to primary consumer is between 5 and 10%
• much of plant body is indigestible, e.g. cellulose and lignin.
• consumer rarely eats whole plant – roots and stems may be left behind. Primary consumers are HERBIVORES – they get their energy by eating plants.

Energy transfer to secondary consumer is between 10 and 20%
• animal material has a higher energy value
• animal material is more digestible. Secondary consumers are CARNIVORES – they get their energy by eating other animals.

Respiration losses occur from each trophic level. Respiration releases the energy necessary to drive metabolic reactions, e.g. active transport and protein synthesis.

PRODUCER

PRIMARY CONSUMER

SECONDARY CONSUMER

Food chains are short because
• long chains would need an enormous producer biomass spread over a vast area
• energy transfer between levels is so inefficient – fox might only receive 10% × 10% × 5% of light energy falling on producer.

The Sun is the source of energy that drives all food chains. Less than 1% of the energy released from the Sun's surface actually falls onto the leaves of the photosynthesizing plants.

Decomposer organisms are fungi and bacteria that obtain their energy and raw materials from plant and animal remains – eventually the entire energy content of these remains will be released as **heat** from inefficient respiration. Some energy may be 'stored' in fossil fuels, e.g. coal, if decomposition cannot be completed.

FOOD WEBS show a network of interconnected food chains. They are called food webs because
• each herbivore usually eats more than one type of producer
• each carnivore must feed on more than one type of herbivore.
The more complex a food web is (i.e. the more food chains involved) the more stable it is, and the less likely it is to be affected by the loss of one type of organism.

HUMANS SHOULD EAT MORE VEGETABLES!
Because of this inefficient energy transfer, it is better for humans to eat green plants and not to waste energy by feeding crop plants to animals.

1. This diagram represents the number of different organisms in a certain food chain.

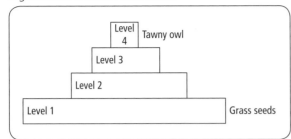

Level 4 — Tawny owl
Level 3
Level 2
Level 1 — Grass seeds

a. What name is given to this type of diagram? (1)
b. Study the diagram and suggest organisms that could be at levels 2 and 3. (2)
c. Why are there fewer organisms at level 3 than level 2? (2)
d. Draw out a similar diagram for the food chain
   Oak tree ⇒ aphid ⇒ blue tit ⇒ sparrowhawk
   Explain why this diagram is not identical to the one shown above. (4)

2. Some students carried out a survey of the animals feeding on an oak tree. They collected a number of different species, and observed which animals fed on leaves and which fed on other animals. The results of their survey are shown in the table below.

ANIMAL COLLECTED/ OBSERVED	NUMBER RECORDED	FOOD EATEN BY ANIMAL
Willow warbler	5	Larvae of winter moths and oak eggars
Winter moth larva	44	Oak leaves
Oak eggar caterpillar	53	Oak leaves
Tawny owl	2	Willow warblers, great tits and field mice
Great tit	5	Larvae of winter moths and oak eggars
Beetle	4	Larvae of winter moths and oak eggars
Field mouse	3	Acorns

Copy the following table. Use the information above to name the animals in the trophic (feeding) levels in this table. (2)

TROPHIC LEVEL	ORGANISMS PRESENT
4: tertiary consumers	
3: secondary consumers	
2: primary consumers	
1: producer	

a. Calculate how many animals there were at each trophic level. Draw a table of your results. (2)
b. Use a piece of graph paper to draw an accurate pyramid of numbers for the animals at the three trophic levels. (4)

c. Why is the number of tertiary consumers not really representative of this particular oak tree ecosystem? (2)

3. A student wished to investigate the population of springtails, a small arthropod, in the soil beneath different types of tree. The teacher suggested using quadrats and a Tullgren funnel, and the investigation was carried out as shown in the diagram below.

Quadrat placed beneath tree

The student took an equal volume of soil from the centre of each quadrat, and then used the Tullgren Funnel to estimate the number of springtails in each sample.

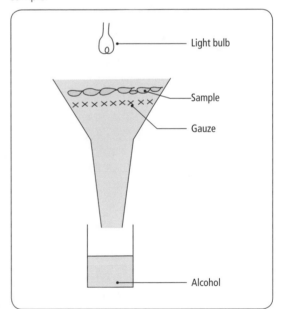

Light bulb
Sample
Gauze
Alcohol

a. The quadrats were distributed in a **random** way. Why was this important? (1)
b. The student used the same-sized sample from each quadrat. Why? (2)
c. The Tullgren funnel has a lamp and wire mesh as important parts. What function do they have? (2)

d. The results of the investigation are shown in the table below.

NUMBER OF SPRINGTAILS IN SAMPLE.	
OAK TREE IN WOODLAND – FALLEN LEAVES BENEATH	OAK TREE IN PARKLAND – FALLEN LEAVES REMOVED
30	10
20	15
45	15
10	5
25	5
25	8
25	12
40	12
15	7
25	11

Calculate the mean springtail numbers in the two different habitats. (2)

e. Oak leaves decompose and release a weak acid. The decomposition process generates heat and water. Decomposing leaves are a food source for many arthropods.

Suggest **one** factor that might be responsible for the different number of springtails in the two habitats. Describe how you would measure this factor. How could you improve the investigation to eliminate the other factors that might affect the springtail numbers? (6)

4. The diagram shows the flow of energy through a food chain. The figures are in kilojoules of energy per square metre per year.

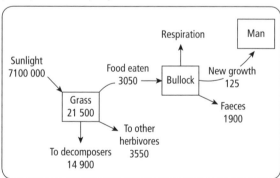

a. How do decomposers break down carbon-containing compounds from the dead remains of grass plants? (3)

b. When the bullock eats the grass, much of the energy from the grass is released in respiration.
   i. How much energy is released by the bullock in respiration?
      .... kJ per m² per year (1)
   ii. Give **one** use of the energy released in respiration. (1)

c. Intensive rearing of cattle indoors is an attempt to reduce energy losses. The table shows the energy balance for indoor and outdoor meat production from cattle.

	kJ PER m² PER YEAR	
	INDOORS	OUTDOORS
Energy input as food	10 000	5 950
Energy input as fossil fuel	6 000	50
Energy trapped in meat	40	1.8

i. The percentage efficiency of rearing cattle indoors is 0.25%. Use the following formula to calculate the percentage efficiency of rearing cattle outdoors.

$$\text{Percentage efficiency} = \frac{\text{Energy trapped in meat}}{\text{Total energy input}} \times 100$$

Show clearly how you work out your answer.
...... Percentage efficiency (2)

ii. Suggest **two** reasons why rearing cattle indoors is more efficient than rearing them outdoors. (2)

iii. Suggest **two** possible disadvantages of rearing cattle indoors. (2)

5. a. A group of pupils were studying a forest. They noticed that the plants grew in two main layers. They called these the tree layer and the ground layer.

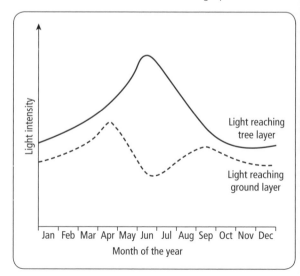

The pupils measured the amount of sunlight reaching each layer at different times in the year. Their results are shown on the graph.

     i.   During which month did most light reach the tree layer?   (1)

    ii.   During which month did most light reach the ground layer?   (1)

    iii. Suggest why the amount of sunlight reaching the ground layer is lower in
        mid-summer than in the spring.   (1)

b. The pupils found bluebells growing in the ground layer.

    Bluebells grow rapidly from bulbs.

    They flower in April and by June their leaves have died.

    i.   Suggest why bluebells grow rapidly in April.   (1)

    ii.   Suggest why the bluebell leaves have died by June.   (1)

c. The pupils hung buckets under the trees and collected the petals and fruit that fell off
    the trees during the year.

    The results are shown on the bar chart.

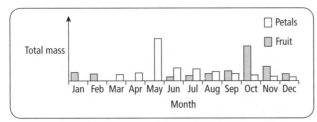

    i.   The trees reproduced by sexual reproduction. Suggest in which month most flowers are fertilized.   (1)

    ii.   The pupils worked out that it took an average of five months for a fruit to grow and be dispersed.
        Suggest how this could be worked out from their bar chart.   (2)

ECOLOGY AND ECOSYSTEMS: Crossword

ACROSS:
3 A unit of energy
5 The living organisms and their non-living environment
6 Piece of apparatus that can separate a piece of the environment for sampling
9 An organism that feeds on the molecules made by producers
10 An organism that must obtain its food molecules ready-made
12 All the living organisms in a habitat
14 A straight-line way of showing a series of feeding relationships
17 An organism that feeds only on plants
18 A part of the environment that can provide food, shelter and breeding sites
19 An organism that feeds on both plant and animal material

DOWN:
1 An organism that must make its own organic compounds from simple raw materials
2 The stage in a food chain that can convert light energy into chemical energy
4 All the organisms of the same species living in a particular area
7 A survey method that is useful for showing changes in living organisms when moving from one habitat to another
8 A meat eater
11 The source of energy for every food chain
13 This element is cycled during respiration and photosynthesis
14 A way of showing interlinked and overlapping food chains
15 Is calculated from number of organisms times mass of each individual
16 A representative part of a population

Cycling of nutrients

involves **interconversion of simple and complex molecules**

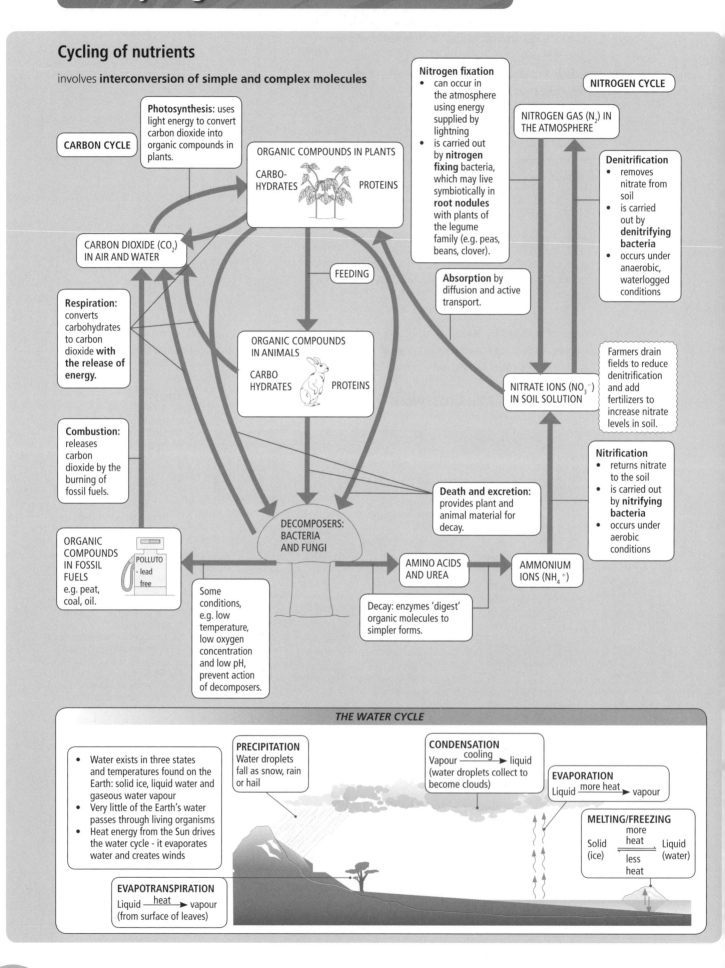

CARBON CYCLE

Photosynthesis: uses light energy to convert carbon dioxide into organic compounds in plants.

ORGANIC COMPOUNDS IN PLANTS

CARBO-HYDRATES   PROTEINS

CARBON DIOXIDE (CO_2) IN AIR AND WATER

Respiration: converts carbohydrates to carbon dioxide **with the release of energy.**

Combustion: releases carbon dioxide by the burning of fossil fuels.

ORGANIC COMPOUNDS IN FOSSIL FUELS e.g. peat, coal, oil.

POLLUTO - lead free

Some conditions, e.g. low temperature, low oxygen concentration and low pH, prevent action of decomposers.

FEEDING

ORGANIC COMPOUNDS IN ANIMALS

CARBO HYDRATES   PROTEINS

DECOMPOSERS: BACTERIA AND FUNGI

AMINO ACIDS AND UREA

Decay: enzymes 'digest' organic molecules to simpler forms.

Death and excretion: provides plant and animal material for decay.

Nitrogen fixation
- can occur in the atmosphere using energy supplied by lightning
- is carried out by **nitrogen fixing** bacteria, which may live symbiotically in **root nodules** with plants of the legume family (e.g. peas, beans, clover).

Absorption by diffusion and active transport.

NITROGEN CYCLE

NITROGEN GAS (N_2) IN THE ATMOSPHERE

Denitrification
- removes nitrate from soil
- is carried out by **denitrifying bacteria**
- occurs under anaerobic, waterlogged conditions

Farmers drain fields to reduce denitrification and add fertilizers to increase nitrate levels in soil.

NITRATE IONS (NO_3^-) IN SOIL SOLUTION

Nitrification
- returns nitrate to the soil
- is carried out by **nitrifying bacteria**
- occurs under aerobic conditions

AMMONIUM IONS (NH_4^+)

THE WATER CYCLE

- Water exists in three states and temperatures found on the Earth: solid ice, liquid water and gaseous water vapour
- Very little of the Earth's water passes through living organisms
- Heat energy from the Sun drives the water cycle - it evaporates water and creates winds

PRECIPITATION Water droplets fall as snow, rain or hail

CONDENSATION Vapour $\xrightarrow{cooling}$ liquid (water droplets collect to become clouds)

EVAPORATION Liquid $\xrightarrow{more\ heat}$ vapour

MELTING/FREEZING Solid (ice) $\underset{less\ heat}{\overset{more\ heat}{\rightleftharpoons}}$ Liquid (water)

EVAPOTRANSPIRATION Liquid $\xrightarrow{heat}$ vapour (from surface of leaves)

1. The carbon cycle is shown below. The arrows represent the various processes that happen in the cycle.

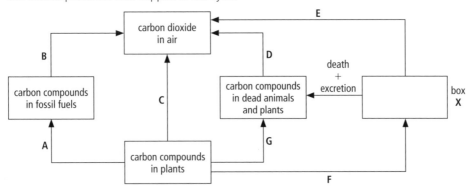

a. i.   Complete the diagram by filling in box **X**. (1)

ii   State the letters of **two** arrows that represent respiration. (2)

iii State the letter of the arrow that can only represent combustion in this cycle. (1)

iv   State the letter of the arrow that represents the process in the cycle that takes millions of years to happen. (1)

b. i   Photosynthesis is not shown on the diagram. Draw an arrow to represent photosynthesis and label it **P**. (1)

ii   Write a word equation for photosynthesis. (2)

CIE 0610 June '07 Paper 2 Q5

2. The diagram below shows the feeding process in a small organism.

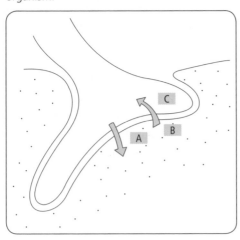

The organism is feeding on some waste protein – the skin of a dead animal.

a.   Which **class** of enzyme must be secreted at A? (1)

b.   Which type of compound will be absorbed at B? (1)

c.   Which Kingdom does this organism belong to? (1)

d.   What is the name given to the feeding structure labelled C? (1)

A biologist was interested in the uptake of the compounds at B. He carried out an experiment to investigate the effect of oxygen concentration on this process. The results are shown in the table below.

OXYGEN CONCENTRATION / %	0	4	8	12	16	20
RATE OF UPTAKE / ARBITRARY UNITS	2	10	18	25	32	32

e.   Draw a graph of this information. (5)

f.   What does the graph tell you about the uptake of compounds at B? (3)

g.   Suggest another factor that might affect the rate of uptake of these compounds. (1)

h.   How could this information be useful to an organic farmer? (2)

3. This diagram shows part of the nitrogen cycle.

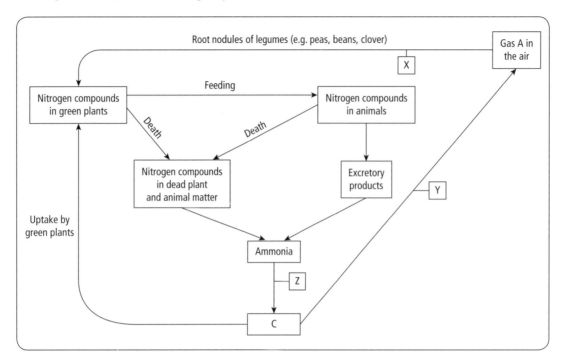

a. Identify the gas labelled A. (1)

b. Many of the stages of the nitrogen cycle depend on the actions of bacteria. Three of these processes are nitrification, nitrogen fixation and denitrification. Match up the labels X, Y and Z with these three processes. (3)

c. Identify the compound labelled C. (1)

d. Name one biological molecule that is made by plants from compound C. (1)

e. Animal excretory products are broken down in the nitrogen cycle. Name one animal excretory product containing nitrogen. (1)

4. Water is essential for living organisms, but there are real worries that farmers' overuse of fertilizers is polluting our water supplies. It is estimated that in the United Kingdom only about half of the fertilizer applied to the soil is actually taken up by plants. Much of it is washed away by rain and drains through the soil.

a. State two ways in which water is used by living organisms. (2)

b. Which ion is most common in fertilisers added by farmers? (1)

c. The concentrations of nitrate and oxygen in a polluted river are shown in the table below.

YEAR	OXYGEN CONCENTRATION / mg dm⁻³	NITRATE CONCENTRATION / mg dm⁻³
1959	5.9	8.4
1964	5.4	10.7
1971	4.5	12.0
1980	4.1	12.9
1988	3.9	13.8

i. Plot a bar chart of this information (5)

ii. Calculate the percentage decrease in oxygen concentration between 1959 and 1988. Show your working. (2)

d. The nitrates and phosphates in fertilizers encourage the growth of simple plants (algae) in the surface layers of the water. A similar effect is caused by the outflow of phosphate detergents. The increased numbers of algae prevent light reaching the lower levels of the water, and rooted plants die. Their remains are broken down by bacteria, which use up large quantities of dissolved oxygen. Fish and other organisms with a high oxygen requirement also die, and are also broken down by microbes, so that the problem gets worse and worse.

i. What name is given to this 'over-feeding' of algae in rivers and lakes? (1)

ii. In unpolluted waters the algae do not grow so quickly. What do you think is limiting their growth? (1)

iii In some developing countries the use of fertilizers on rice paddies has affected local lakes and rivers. Fish have died or have migrated to less-polluted waters.

What important nutrient for humans would fish supply in a developing country? (1)

CYCLING OF NUTRIENTS: Crossword

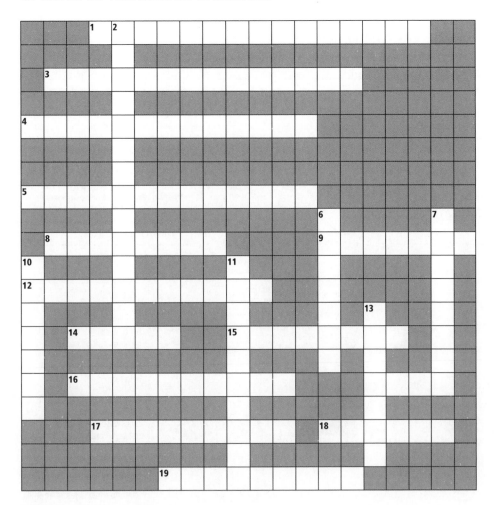

ACROSS:
1 Loss of nitrogen from nitrogen compounds to the atmosphere
3 The process that converts 4 across to carbohydrates
4 The most common carbon-containing gas (6,7)
5 Conversion of ammonium salts to nitrate
8 Process that makes nitrogen gas available as nitrogen-containing compounds
9 Underground homes of bacteria responsible for 8 across
12 Releases 4 across as foods are oxidized
14 Decomposers that have bodies made of hyphae
15 Microbes responsible for much of the cycling of nutrients
16 An organism that breaks down complex molecules into simpler ones
17 One way of returning animal wastes to the environment
18 This type of transport is needed for the uptake of many minerals
19 Simple nitrogen-containing compound in living organisms (5,4)

DOWN:
2 The over-feeding of lakes and ponds with excess nitrates
6 Molecules needed by 16 across to break down large molecules
7 The washing of nitrates from the soil into water
10 Complex organic compounds: contain nitrogen and are needed for growth
11 Process that releases 4 across from fossil fuels
13 Important ionic form of nitrogen

Chapter 35:

Human impact on the environment

Impact of humans on the environment
We now know that human activities can have a major impact on the environment.

GLOBAL WARMING
Greenhouse gases (e.g. carbon dioxide and methane) keep heat in and reflect it back to Earth. Greenhouse gases from industry may be causing global warming. This could cause:
- greater climatic extremes (high winds and heavy storms)
- melting of ice taps, causing seas to rise and flood low-lying land
- greater evaporation from soils, causing crop losses

NUCLEAR FALL OUT
Ionizing radiation (α and β particles and γ rays) can be very damaging
- because it can increase mutation rates, which could cause cancer including **leukaemia**
- the sources of radiation take many years to decay (it may have a **long half-life**).

The two ways in which the atmosphere may be polluted by nuclear fall out are
- from nuclear power plant accidents
- from nuclear weapons.

DAMAGE CAUSED BY EXCESSIVE USE OF PESTICIDES
Some pesticides cause damage when they enter the food chain because the chemicals accumulate in the bodies of the higher predators.

Accumulation of pesticide
pesticide in water microscopic
water plants small animals,
e.g. water fleas
small fish
larger fish
grebe (fish-eating
water fowl).

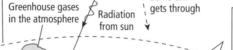

Ozone layer

U.V. radiation

Some u.v. gets through

Greenhouse gases in the atmosphere

Radiation from sun

'Heat'

ACID RAIN
Combustion of fossil fuels in industry, homes, and car engines produces sulfur dioxide and gaseous oxides of nitrogen. When these dissolve in water, acids are formed. These fall as acid rain.

Acid rain makes the soil acid. This damages plant roots and takes essential minerals from the soil. Poisonous chemicals are also released. Plants and trees may be killed. Acid rain and chemicals washed from the soil enter rivers and lakes. This kills fish and other aquatic animals.

POLLUTION FROM SEWAGE AND FERTILIZERS
Sewage and fertilizers from farmland on enter lakes and rivers. The nutrients (nitrates and phosphates) make the algae in the water grow rapidly. Other plants are smothered and die.

Bacteria use up oxygen from the water and so fish and other animals die.

This process is called **eutrophication**.

HABITAT DESTRUCTION
Clearing woodlands, forests, swamps and wetlands in order to create more land for farming or building may destroy the natural habitats of rare plants and animals. Some may be lost forever.

Deforestation also increases soil erosion by removing tree roots that held the soil. In same countries this leads to flooding and mud slides.

1. A school party was spending some time on a rocky shore. Some of the students decided to make a survey of the different molluscs on the shore. The following diagram represents a section of the shore that had been sampled taking line transects across it.

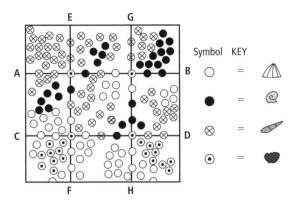

a. Which of the four transects corresponds to the following diagram? (2)

b. From the results of this transect alone, which mollusc type seems to be the most abundant? Give its symbol. (2)

c. Use the diagram and your answer to **b** to suggest why line transects sometimes give unreliable data. (2)

d. The section of shore shown below measures 10 metres by 10 metres. Estimate the total number of limpets on this section of shore **using only the information provided by the random quadrats**. (3)

Quadrats in randomly chosen sites
(each quadrat encloses one square metre)

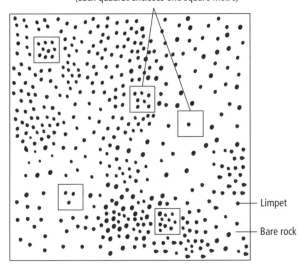

Limpet

Bare rock

e. Oil was released from the tanks of a passing ship, and some washed up on this section of shore. The diagram below shows this section of shore four weeks after the oilspill. Using the same method, estimate the **new** total of limpets. (3)

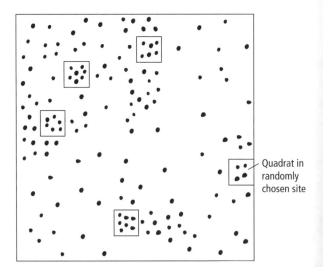

Quadrat in randomly chosen site

f. **From your results**, does the oil seem to have affected the limpet population? Explain your answer. (2)

g. **From an overall view of the two diagrams**, does the oil seem to have affected the limpet population? Suggest a source of error in the sampling technique that was responsible for failing to show up the effects of the oil. (2)

2. The diagram below shows the sequence of events associated with acid rain pollution. Complete the diagram using terms chosen from the following list.

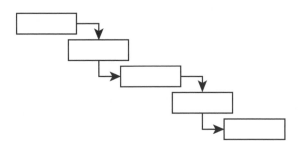

A: sulfur dioxide rises into the atmosphere
B: acid rain falls
C: fish die in acidified lakes
D: discarded car batteries leak acid
E: fuel combustion in power stations
F: ozone is produced in power stations
G: pollutant combines with water vapour

3. Householders in the UK produce approximately 20 million tonnes of domestic waste each year. The table shows the average composition of domestic waste.

TYPE OF WASTE MATERIAL	PERCENTAGE OF TOTAL
Paper and board	32
Food and garden rubbish	24
Glass	10
Plastics	7
Metal	4
Other	

    a. Complete the table, and then draw a bar graph of the information it contains. (5)

    b. What is meant by the term **biodegradable**? Give one example of material that is biodegradable. (3)

    c. Much food and garden rubbish is disposed of in plastic bags. The interior of the bag is anaerobic, and an explosive gas can be formed. What is the name of this explosive gas? (1)

    d. If the 'other' category is ignored, what percentage of the total domestic waste could be biodegradable? Show your working. (3)

4. a. Draw a simple diagram to explain the meaning of the term **global warming**. (3)

    b. Name **two** different greenhouse gases. (2)

    c. Explain how global warming might affect

      i. sea level

      ii. the pattern of insect pest distribution around the world

      iii. the likelihood of gales and storms (3 × 2)

5. a. i. Which form of the Sun's energy is used by plants? (1)

      ii. Name the process that uses this absorbed energy. (1)

    b. The graph shows how the concentration of carbon dioxide in the atmosphere has changed over a period of about 20 years.

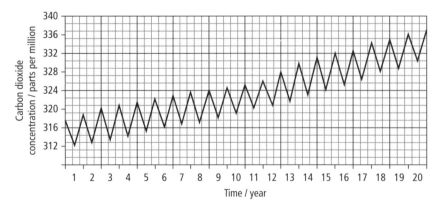

        Describe the changes shown by this graph. (2)

    c. The atmosphere around the earth acts as a trap for energy from the Sun.

      Carbon dioxide in the air traps heat energy.

      i. Suggest the effect the overall change in the graph may be having on the Earth's climate. Explain your answer. (3)

      ii. Humans cause changes in ecosystems, including changing the amount of carbon dioxide in the atmosphere.

      Suggest two ways In which the overall change can be reversed. (2)

CIE 0610 November '05 Paper 2 Q5

6. Deforestation occurs in many parts of the world.

    a. State two reasons why deforestation is carried out. (2)

    b. i. Explain two effects deforestation can have on the **carbon cycle**. (4)

      ii. Describe two effects deforestation can have on the **soil**. (2)

      iii. Forests are important and complex ecosystems. State **two** likely effects of deforestation on the forest ecosystem. (2)

CIE 0610 June '06 Paper 2 Q2

REVISION SUMMARY: Fill in the gaps

Complete the following paragraphs about pollution. Use terms from the following list – you may use each term once, more than once or not at all.

PESTS, CFCs, ULTRAVIOLET LIGHT, NITRATE, BATTERIES, MESOPHYLL, CARBON DIOXIDE, ACID RAIN, INFRA-RED, METHANE, NITROGEN, PESTICIDES, CANCER, PHOTOSYNTHESIS, MUTATION, MOTOR OIL, OZONE, OXYGEN.

The greenhouse effect may cause global warming when .... radiation is trapped close to the Earth's surface by a layer of .... from ruminants, ... from aerosols and .... from the combustion of fossil fuels. Global warming may allow .... to extend their range, but may have the benefit of increasing production of food by ....

Holes in the ozone layer may result from ...., formerly used as refrigerants. These holes allow the entry of too much .... radiation, which may lead to skin .... and an increased rate of .... (especially dangerous as these may be passed on to the offspring of the affected person). The production of excess .... may also be damaging – for example, photosynthesis is reduced when the leaf .... is damaged.

Humans are also responsible for pollution of water, for example with excess .... washed out of farmland and with .... used to control theaquatic stages of pests. Atmospheric pollutants can fall as .... and so lower the pH of bodies of water – this can seriously reduce the suitability of the water for living organisms.

Pollution of land is often a result of the inefficient disposal of waste, such as ...., which contain cadmium, ..., which contains toxins that may cause skin cancer and old refrigerators, which contain .... (18)

Agriculture: Pluses and minuses

DEFORESTATION
Removal of trees
+ • provides more land for crops
  • removes habitat for pest herbivores … but …
− • removes habitat for other species
  • can affect water cycle
  • leads to rapid erosion of soil

IRRIGATION SYSTEMS
+ • provide water for growing plants, therefore removing a LIMITING FACTOR … but …
− • often withdraw water from natural sources
  • can lead to LEACHING of minerals from the soil

USE OF FERTILIZERS
The availability of mineral nutrients is an important LIMITING FACTOR for crop growth. Especially
+ • NITRATE for proteins and nucleic acids: leaves
  • PHOSPHATE for nucleic acids, ATP and membranes: flowers and fruits
− • MAGNESIUM for chlorophyll: leaves and stems
So farmers ADD MORE FERTILIZERS … but …
EXCESS FERTILIZERS run off into ponds and streams

EUTROPHICATION

MECHANISATION
Use of tractors and harvesting machinery
+ • means cultivation can be quicker, and more land can be used … but …
− • more fossil fuels are used, and more pollution results
  • soil is compacted, so it becomes more difficult for rainwater to penetrate

NPK
NPK

MONOCULTURE
Growing a single crop in one habitat
+ • nutrients and pest control can be exactly matched to crop
  • mechanical harvesting is easier … but ..
− • outbreaks of pests and diseases can spread very rapidly
  • loss of food chains reduces biodiversity
  • removes some nutrients specifically

MIXED CROP ROTATION
• reduces pest infestations
• improves nutrient balance
✓

USE OF PESTICIDES
+ less competition for nutrients, light and water … but …
− pesticides accumulate in food chains, with top predators being particularly affected

DDT

REMOVAL OF HEDGEROWS
+ • more growing space for crops
  • less competition for light, water and nutrients
  • fewer habitats for pests… … but …

NO NEST!
GONE

− Less shelter and fewer nesting sites for birds, pollinating insects and beetles, which are predators on pests

Managing ecosystems: animal husbandry

Intensive farming aims to minimize productivity by reducing environmental resistance.

Temperature control: essential so that costly heat energy is not wasted.
If **too high**, animals are uncomfortable and will not feed
If **too low**, food intake is 'wasted' on heat production to maintain body temperature.

Controlled light regime/ photoperiod: may influence growth rate because it can permit longer feeding period; may control reproductive cycle so, for example, milk production may be stabilized through the year.

Shelter:
Prevent entry of predators. Eliminate/control **competitors** for food. Protect against **climatic extremes**.

Food input:
Control content - high protein for growth
    - minimal fat to suit customer demand for lean meat
    - include growth hormone to increase growth rate
    - add copper ions which reduce energy consumption for heat production
Often use dried milk powder or single cell protein, with mineral and vitamin supplements.

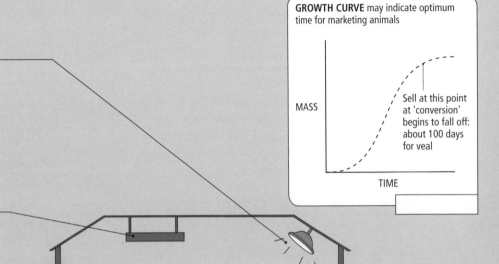

GROWTH CURVE may indicate optimum time for marketing animals

MASS / TIME

Sell at this point at 'conversion' begins to fall off: about 100 days for veal

Strain of animal: selective breeding for animals with high conversion ratio, i.e. most efficient transfer of food intake to body mass.
Minimise movement: less energy consumption and thus more efficient 'conversion'.
NB can upset **social interactions**, and stress can lead to poor growth/ unpleasant 'cortisone' taste.

Hygienic conditions: most animals are free of gut parasites and debilitating bacteria – healthy animals grow more quickly and meat is more saleable.
Slurry (faeces/urine) can be dried and recycled for use as fertilizer.

Veterinary care:
Antibiotics to reduce bacterial infections
Vaccination to minimize viral infections
Hormone/vitamin supplements can be administered more accurately than in the diet
Artificial insemination techniques can reduce costs (no need to keep bulls in dairy farms).

1. In order to increase the yield of crops, many farmers have used some or all of the following practices:
    a. cutting down hedges
    b. increased use of nitrate fertilizers
    c. burning of stubble after harvesting
    d. drainage of wet fields
    e. repeated growth of the same crop in the same field
    f. deep ploughing of soil using heavy machinery

    Many conservationists believe that these techniques may be damaging to the environment. Choose any **two** of the above and state how they help the farmer to increase the yield of crop, then choose **three** other techniques and explain how they might damage the environment.
    Write your answers in a table like this one. (5)

TECHNIQUE	BENEFIT TO FARMER

TECHNIQUE	HARM TO THE ENVIRONMENT

2. The following diagrams represent population distributions in two different countries.

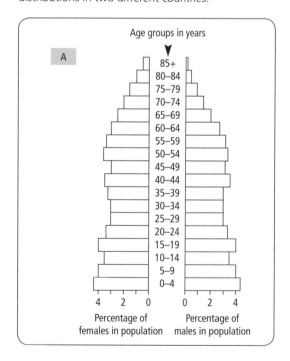

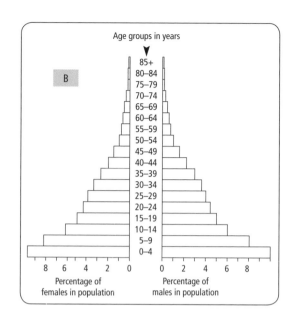

a. In age pyramid A, what percentage of the male population are aged 60 and over? (1)
b. In age pyramid A, which sex has the higher proportion reaching old age (65+)? (1)
c. In age pyramid B, what proportion of the population are female aged 15–19? (1)
d. In age pyramid B, which age group makes up exactly 4 per cent of the male population? (1)
e. Which age pyramid represents the population of a developing country? Give **two** reasons for your answer. (2)
f. In the country represented by age pyramid A there is a population of 60 000 000. How many are males aged 15–19? (2)
g. Explain how the agricultural revolution allowed an increase in population of the United Kingdom. (3)

3   a. The table below shows one possible system of crop rotation used in agriculture.

FIELD	YEAR 1	YEAR 2	YEAR 3	YEAR 4
A	Sprouts	Peas	Potatoes	Fallow or cereal
B	Fallow or cereal	Sprouts	Peas	Potatoes
C	Peas	Potatoes	Fallow or cereal	Sprouts

i. Write out the likely sequence of a fourth field, field D. (1)
ii. What is meant by the term 'crop rotation'? (2)

b. The level of nitrate in the soil can be measured. The values obtained for the sprout and the pea fields are shown in the following table.

CROP	FIELD	NITRATE LEVEL IN SOIL / ARBITRARY UNITS
Peas	A	82
Peas	B	88
Peas	C	80
Sprouts	A	54
Sprouts	B	44
Sprouts	C	52

i.   Draw a bar chart of these results.   (3)
ii. Calculate the percentage increase in nitrate level when peas are grown. Show your working.   (3)

4. Study the revision diagram on animal husbandry and answer the following questions.
a. Why are the abiotic factors of **temperature** and **light** so important to maximizing yield?   (2)
b. Suggest **three** features of the food pellets used to feed the animals.   (3)
c. Why is it important to weigh the animal at regular intervals?   (2)
d. How could liquid slurry have a bad effect on the environment if it was allowed to run into nearby rivers?   (3)
e. What are the benefits to productivity of regular veterinary care?   (2)

REVISION SUMMARY: Match the terms with their definitions

	TERM			DEFINITION
A	Age pyramid		A	A mineral nutrient added to increase crop yield
B	Biodiversity		B	The cultivation of a single type of crop plant
C	Crop rotation		C	Plant type able to host bacteria capable of nitrogen fixation
D	Eutrophication		D	Some resource in the environment that reduces the growth of an individual or population
E	Fertilizer		E	A way of showing how many members of a population belong in each age group
F	Legume		F	The amount of growth (biomass added) that takes place over a fixed period of time
G	Limiting factor		G	Growth of algae in ponds and rivers due to 'overfeeding' with nitrate and phosphate
H	Monoculture		H	A chemical used to control the growth and reproduction of organisms that compete for human crops
I	Pesticide		I	The range of different species of living organisms
J	Productivity		J	The regular changing of the type of food plant grown in a particular field over a period of years

Chapter 37: Conservation

Conservation of species

There are several possible strategies

PRESERVATION: involves keeping some part of the environment **without any change**. Only possible if the area can be fenced off / protected

RECLAMATION: involves the restoration of **damaged** habitats. Often applies to the recovery of former industrial sites, such as mine workings

CREATION: involves the production of **new** habitats. Only really possible with small areas, e.g. digging a garden pond, or planting of a new forest

ALL INVOLVE MANAGEMENT OF A HABITAT

DIRECT PROTECTION MAY BE NECESSARY

Even when a suitable **habitat** (able to provide food, **shelter** and **breeding sites**) is available, individual species may be at direct risk from humans.

- Rhinoceroses may be hunted for their horns, mistakenly believed to have medicinal or aphrodisiac properties.
- Elephants may be hunted for their ivory tusks.
- Primates (e.g. chimpanzees) and other species may be hunted as 'bush meat'
- Butterflies, molluscs (shells) and plants may be 'collected'
- Parrots, primates and fish may be collected for the pet trade

ONE USEFUL TECHNIQUE INVOLVES...

PRESSURES ON A HABITAT

- Humans have a significant **biotic** impact on the environment

  Temporary when humans were **nomadic**: the environment has periods of recovery

  Permanent as humans became **cultivators** and **settlers**

- Use of tools and domestication of animals → more efficient agriculture / support for larger population
- Greater demands for shelter, agricultural land and fuel → greater rate of deforestation
- Development of fossil fuels → more use of machines / greater 'cropping' / larger populations
- Pollution as greater use of fossil fuels / pesticides / fertilizers and development of nuclear power

FLAGSHIP SPECIES

Large, attractive and 'cuddly' species attract funding from agencies and donations from the public protection for less attractive species (e.g. beetles and worms) that live in the same habitat.

MANAGEMENT IS A COMPROMISE!

- Maintenance of a particular habitat (e.g. chalk hillside for wild flowers / butterflies) often means **halting succession**.
- The requirements for wildlife must be balanced by Human demands for **resources** (e.g. mining for uranium), **recreation** (e.g. diving around coral reefs) and **agricultural** land.

1.

DDT AND HERONS IN UGANDA

Nine clutches of eggs from the green heron were collected close to the Kasinga Channel in Southern Uganda, during 2002. The sites were close to sample sites checked in 1990. Analysis of the eggs showed that they had an average increase in DDT concentration of 60 per cent, and the eggshells were nearly 30 per cent thinner compared with 1990. The lilac breasted roller, which lives in drier, cattle-rearing areas, has seriously declined in numbers over the same period: this bird feeds on insects taken from tree trunks in areas where DDT is used to control tsetse flies.

a. What was the average annual increase in DDT concentration in the Green Heron eggs? Show your working. (2)

b. Why is it difficult to estimate the effects on the whole population from the data available? (1)

c. Explain why high levels of DDT are thought to reduce breeding success of these birds. (2)

d. Explain why the Green Heron is likely to take up high levels of DDT. (2)

e. Suggest how the feeding method of the roller make it particularly at risk from DDT (1)

f. In the United Kingdom, DDT use was blamed for a 60 per cent reduction in the number of sparrowhawks in the 1960s. By 1990 sparrowhawk numbers had almost recovered. Explain why sparrowhawk numbers recovered in the UK, but Green Heron numbers in Uganda are less likely to recover. (2)

2. Toads are amphibians. Only two species are native to Britain, the Common toad (*Bufo bufo*) and the Natterjack toad (*Bufo calamita*).

Natterjack toads like warm sandy soil in open and sunny habitats, with shallow pools for breeding. Examples of these habitats are heathland and sand dunes. Common toads like cooler, more shady habitats, such as woodland.

Many areas of sand dunes are being developed for camp sites. Heathland can easily change to woodland as trees grow on it. In the summer, woodland is colder than heathland due to the shade the trees create.

These conditions suit the Common toad, but not the Natterjack. As a result of the changing habitats the Natterjack toad is becoming an endangered species.

a. i.   Name **one** external feature that identifies an animal as an amphibian. (1)
    ii. Amphibians are a class of vertebrate. Name two other vertebrate classes. (2)
b. State **one** piece of information from the passage to show that the Common toad and Natterjack toad are closely related species. (1)
c. From the information provided, state two reasons why Natterjack toads are becoming endangered. (2)
d. Suggest measures that could be taken to protect the Natterjack toad from extinction. (2)

Below is a food web for British toads.

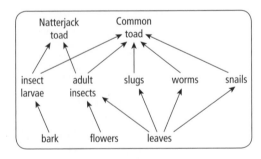

e. i.   State the trophic level of toads. (1)
    ii. State which foods the two species of toad both eat. (1)
    iii. With reference **only** to food, suggest why the Common toad is more likely to survive when the two species are in competition. (1)

CIE 0610 November '05 Paper 3 Q1

3. The Ruddy duck, *Oxyura jamaicensis*, is a native of America.

A flock of 20 birds was introduced into Britain from America before 1950. The original flock settled quickly in their new habitat and started breeding. Numbers now exceed 6000.

The White-headed duck, *Oxyura leucocephala*, (a native of Spain) is a closely related species to the Ruddy duck.

Female White-headed ducks are more attracted to mate Ruddy ducks than to males of their own species. Cross-breeding between the two species produces a new variety of fertile duck.

The White-headed duck is now threatened with extinction. Some conservationists are considering a plan to kill the British population of Ruddy ducks to prevent the White-headed duck becoming extinct.

The diagram below shows a male Ruddy duck.

a. State two features, visible in the diagram, that distinguish birds, such as the Ruddy duck, from other vertebrate groups. (2)
b. i.   With reference to an example from the passage, describe what is meant by the term *binomial system*. (2)
    ii. State two reasons, based on information in the passage, why the Ruddy duck and White-headed duck are considered to be closely related. (2)
c. i.   Explain why Ruddy ducks would **not** become extinct, even if British conservationists carried out their plan. (1)
    ii. Suggest **one** factor, other than the breeding habits of the Ruddy duck, that could result in the extinction of a bird such as the White-headed duck. (1)
d. The Ruddy duck feeds on seeds and insect larvae. The ducks are eaten by foxes and humans.
    Explain why these feeding relationships can be displayed in a food web, but not in a food chain. (2)

CIE 0610 June '05 Paper 3 Q6

4.

GOLDEN LION TAMARINS AND ZOOS

Golden lion tamarins are small South American primates, living in the coastal forests of Brazil. They require extensive forests of tall trees as they have to range widely to find food, and must avoid predators both on the ground and from the air. Many of these forests have been cut down for timber and for cattle ranching.

They breed slowly and have small litters. The mortality rate among the young animals is high, mainly because of predation.

Golden lion tamarins have been kept in zoos for a number of years but there are problems with the captive breeding programme: many of the offspring are genetically related to one another, causing a reduction in breeding rate, and adults cannot learn the tree-climbing and feeding skills they need when confined to zoos

These animals are very attractive and suffer from poaching for their skins, even though trade in these skins is banned by CITES.

a. Give two features of the Tamarin that would confirm that it is a mammal. (2)
b. Why is the Golden Lion Tamarin in danger of becoming extinct? (4)
c. Give three arguments against keeping Golden Lion Tamarins in zoos. (3)
d. Give two arguments in favour of keeping Golden Lion Tamarins in zoos. (2)
e. What part does the organization CITES play in the conservation of species? (2)
f. The Golden Lion Tamarin is a 'flagship species'. What does this mean? (2)

5. Read the following article, and then answer the questions that follow it.

Following the Second World War the British government made plans to increase crop yields in Britain. Between 1945 and 1965 hedgerows were being removed at a rate of 3000 miles per year – this rate of removal increased to 5000 miles per year by 1968. The rate of removal has fallen recently, and a recent report suggests that between 1980 and 1990, 10 000 miles were removed but 5000 miles were replanted. Conservationists argue that new hedgerows are a much poorer habitat for wildlife than older, mature hedges.

In Cambridgeshire, farmers estimated that the removal of one mile of hedgerow provided an extra three acres of arable land and reduced by 40 per cent the time taken to harvest a field of cereals.

The British government spends over £2 billion per year to buy and store surplus food produced in the UK. Much of the food is eventually destroyed, or given away at very low prices. The European Union is now encouraging farmers to produce less food.

a. What was the highest rate of hedge removal before 1969? (1)
b. Why is it less important now to gain extra arable land than it was immediately after the Second World War? (2)
c. English Nature, an important conservation organization, publishes a list of important points about hedgerows. These are some of the points included in the list:
    i.   hedges obstruct the efficient use of modern farm machinery
    ii. hedges are an attractive feature of the rural landscape
    iii. hedges provide cover for game birds such as pheasant, partridge and quail
    iv. hedges can provide a habitat for weeds, rabbits and insect pests
    v.   hedges are very expensive to maintain, barbed wire is very cheap to maintain
    vi. hedges shade part of the crop, and may compete for water
    vii. hedges prevent wind damage to crops
    viii. hedges provide an important habitat for insects that can help with pollination and with biological control of pests

Choose **three** statements from this list that would support hedge removal, and **three** points that could be used to support arguments against hedge removal.

Three statements that support hedge removal are ....

Three statements that argue against hedge removal are ....   (6)

CONSERVATION: Crossword

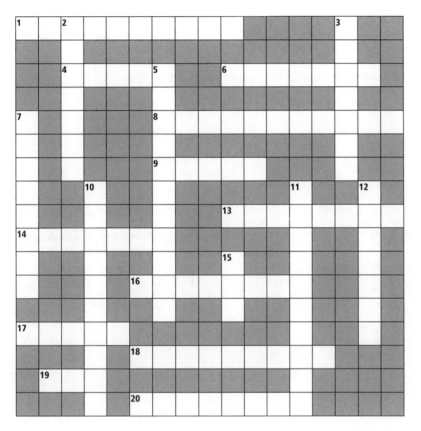

ACROSS:
1 The use, by humans, of a habitat for leisure purposes
4 Symbol of the Worldwide Fund for Nature – an important example of 7 down
6 Interlinked set of feeding relationships – easily disturbed if one species is removed (4,3)
8 Returning a damaged habitat to its original purpose – an important type of conservation
9 Abbreviation for the international organization that bans the trade in endangered species
13 A type of conservation in which new habitats are produced
14 A part of the environment that can provide food, shelter and breeding sites for living organisms
16 With 2 down – one method for increasing the population of an endangered species
17 Large member of the cat family that is endangered because some forms of medicine believe its body parts can treat a range of diseases
18 Humans may damage a habitat as they seek these
19 A place where animals may be protected, observed and where 2 down may take place
20 Large herbivore hunted for the ivory trade

DOWN:
2 See 16 across
3 With 7 down – an important organism in conservation as it has enough public appeal to help raise funds
5 As humans became more efficient at this, more land was taken away from wildlife
7 See 3 down
10 Large herbivore, endangered because some people believe that its horn has aphrodisiac properties
11 All conservation involves this – this is how humans, especially scientists, can be useful in conservation
12 Type of animals collected for its beautiful shell
15 Powerful insecticide that can become concentrated in a food chain and reduce the breeding success of top carnivores

Index